Grok: The Complete Guide

xAI's Real-Time Assistant — Grok 4.3, DeepSearch, Aurora, and Live Knowledge

Ken Tannenbaum

2026

Finnoybu Press

Grok: The Complete Guide

2026 Edition

xAI's Real-Time Assistant — Grok 4.3, DeepSearch, Aurora, and Live Knowledge

---

Published by Finnoybu Press

For permissions and inquiries: press@finnoybu.org

First Edition: 2026

About the Author

Ken Tannenbaum is a technologist, entrepreneur, and AI governance researcher. He is the founder of Finnoybu Press and the founder/steward of the AEGIS Initiative, an open framework for AI safety and accountability. Learn more at aegis-initiative.com.

When he's not building governance standards for artificial intelligence, he's helping everyday people understand and use it.

Dedication

For everyone who wanted an assistant that knows what happened this morning, not just what was true at training time.

The AI for Everyone Series

This book is part of the **AI for Everyone** series by Finnoybu Press, designed to help real people use AI tools effectively. Visit **press.finnoybu.org** for the latest titles and updates.

Collection One: Product Deep Dives

- *ChatGPT: The Definitive Guide (2026 Edition)*
- *ChatGPT: The Power User Guide (2026 Edition)*
- *Claude: The Definitive Guide (2026 Edition)*
- *Claude Code: The Power User Guide (2026 Edition)*
- *Gemini: The Complete Guide (2026 Edition)*
- *Gemini: The Advanced Guide (2026 Edition)*
- *Copilot: The Workplace AI Guide (2026 Edition)*
- *Copilot: The Advanced Workplace Guide (2026 Edition)*
- *Perplexity: The Research Playbook (2026 Edition)*
- *Perplexity: The Advanced Research Playbook (2026 Edition)*

Collection Two: Product Guides

- *NotebookLM: From Sources to Audio Overviews (2026 Edition)*
- *DeepSeek: Code, Math, and Reasoning (2026 Edition)*
- *Cursor: The AI Code Editor (2026 Edition)*
- *OpenClaw: Agentic AI for Everyday Work (2026 Edition)*

Collection Three: Cross-Platform Skills

- *AI Systems Playbook (2026 Edition)*
- *AI Prompt Engineering (2026 Edition)*
- *AI for Developers (2026 Edition)*
- *AI for Business (2026 Edition)*
- *AI for Students (2026 Edition)*
- *AI Governance for Practitioners (2026 Edition)*
- *Securing Agentic AI (2026 Edition)*

The Extended Series — Product Guides, Volume II

- *Codex: The Cloud Coding Agent (2026 Edition)*
- *Grok: Real-Time AI from xAI (2026 Edition)*
- *Mistral: Vibe and the Open European AI (2026 Edition)*
- *DeepL: Translation, Writing, and Voice (2026 Edition)*

The Extended Series — The Capability Set

- *Automate Your Work with AI (2026 Edition)*
- *Build Apps Without Code (2026 Edition)*
- *AI Literacy for the Workplace (2026 Edition)*

Table of Contents

Introduction

I wrote this book because most AI assistants live in the past, and Grok is the one that made the alternative obvious.

For the first few years of the assistant era, every one of these tools shared a quiet limitation we mostly stopped noticing: they were trained up to a date and then frozen. Brilliant about everything before the cutoff, blank about everything after. You could ask the best model in the world what happened yesterday and it would tell you, politely, that it did not know. We adapted — we learned not to ask about anything recent — and the blind spot became so ordinary it stopped registering as a blind spot at all.

Then Grok treated the present as a question it could actually go and answer. Built by xAI inside Elon Musk's X, it has native, real-time access to the live firehose of human conversation and current events, and it reaches into that stream by default rather than apologizing for not knowing. Ask it what is happening right now, what people are saying about something this afternoon, whether a thing has actually shipped, what the live reaction to an event has been — and it goes and looks, and tells you, with sources you can click. That is a different job than its frozen predecessors do. Not better at everything; different in a way that turns out to matter more than any benchmark.

I want to be careful, right at the start, about what kind of book this is and is not. It is not a brochure. Grok is made by a company with a famously outspoken founder, it carries a "less filtered" reputation that is partly earned, and it has been at the center of more than one public controversy. A book that waved all of that away in order to sell you on the tool would not be worth your time, and a book that led with the controversy as though it were the whole story would be just as useless in the other direction. What I have tried to write is a working guide: an honest account of what Grok is genuinely good at, where it is merely competitive, where it is behind, and where it asks for judgment you should not outsource to it. The real-time strength is real and I will make its case without apology. The risks that come bundled with it are also real, and I will not bury them in a final chapter you might skip.

A word, too, about who I imagine on the other side of the page. I have assumed you are busy, a little skeptical, and uninterested in hype — that you have probably already used at least one AI assistant, formed opinions about it, and want to know whether Grok earns a place in your kit rather than whether it is "the best." That is the right question, and it has a real answer that depends on the work you actually do. Much of this book is built to help you find your own answer rather than to hand you mine.

What This Book Covers, Briefly

Grok is xAI's frontier AI assistant — Grok 4.3 is the flagship as of this writing — that holds a real conversation, reasons carefully when a problem is hard, writes and analyzes and codes, sees images and documents, generates pictures through Aurora and short video through Grok Imagine, talks with you in a voice mode, and, distinctively, draws on real-time knowledge from X and the live web to tell you what is true *now*.

This book covers all of it: getting access and choosing among the tiers, reaching Grok on X and the web and your phone, the real-time research that is its signature strength, its creative tools, putting it to work, building on its API — and, just as honestly, where it goes wrong, the "less filtered" reputation it carries, what happens to your data, and how to use it responsibly at work. It also places Grok squarely beside the assistants you may already use — ChatGPT, Claude, Gemini — with a clear account of where it leads and where it does not.

Who This Book Is For

- **Individuals** who want a capable everyday assistant that also happens to know what is going on right now.
- **Professionals and teams** who can use the real-time edge for something specific — monitoring, research that has to include the last day, analysis where being current is the whole value.
- **People comparing assistants**, deciding whether Grok belongs in their kit alongside or instead of what they already pay for.
- **Developers and builders** who want to put Grok's sense of the present inside their own software.

You do not need to be technical to get most of this book. The developer material in Part VI is signposted so you can navigate around it, and everything before it assumes only curiosity and basic comfort with the tools you already use.

How to Read This Book

The book is in seven parts, plus reference.

Part I: Meet Grok is orientation — what Grok is and isn't, where xAI and Grok came from, how it compares to its rivals, and what the Grok 4.x models under the hood actually are.

Part II: Getting Started is the practical setup — the access tiers (read Chapter 5 before you spend anything), where you reach Grok, your first conversations, and the Think mode that powers its harder reasoning.

Part III: Real-Time and Search is the heart of the book — DeepSearch and DeeperSearch, the live knowledge from X, and the research workflows that turn the real-time advantage into real work.

Part IV: Creating with Grok covers the studio side — Aurora for images, Grok Imagine for video, voice mode, and vision for reading what you show it.

Part V: Grok for Work is the professional craft — writing, analysis, coding, and everyday productivity, with the review discipline each one demands.

Part VI: Building on Grok is for developers — the API, tool use and the standout live-search tools, and managing cost.

Part VII: Judgment and Limits is the honest close — what Grok gets wrong, the bias-and-tone question, privacy and your X data, and responsible use at work.

Reference wraps with a model-and-pricing snapshot, an API quickstart, a glossary, and resources for staying current.

Read straight through or jump around. New to Grok: Parts I and II, then III. Here for the real-time strength specifically: Part III. Evaluating it against what you use: Chapter 3, then Part VII. Building on it: Part VI. Anyone bringing it to work: Part VII, end to end.

If you want the shortest possible on-ramp, here it is. Read this introduction and Chapter 1 to get the shape of the tool. Read Chapter 5 before you spend a cent, because Grok's access tiers are genuinely confusing and the worst way to start is by paying for the wrong thing. Then do one real piece of your own work in Grok — not a toy prompt, but an actual question you needed answered today — and pay attention to where it surprised you in either direction. That single hour of honest use will teach you more about whether Grok belongs in your kit than any amount of reading, this book included. Everything else here is in service of making that hour, and the hundred that might follow it, more productive and less likely to go wrong.

A Note About Accuracy

This corner of the AI market moves in weeks, not years. Grok ships changes constantly; model names, prices, tier boundaries, features, and interface details will have drifted by the time you read this. This book is accurate as of mid-2026.

The durable parts — what real-time AI is for, how to use the live edge without being fooled by it, where to trust the tool and where not to — will hold. The screenshots and the price points will not. Where a detail is the kind that changes fast, I flag it and point you at xAI's current documentation rather than pretending the number is permanent. This is true with extra force for pricing, which is why Chapter 5 asks you to confirm the current figures with your own eyes.

A Note About Cost and Access

Grok is not entirely free, and the most powerful parts of it sit behind paid tiers — reachable through either an X subscription or xAI's standalone plans, which is a genuine source of confusion the access chapter untangles. There is a free tier worth starting on, and there is real value gated above it. I am explicit throughout about which tier a capability needs, because few things waste a reader's time like a chapter spent on something they cannot reach. Chapter 5 maps the landscape before you spend anything.

Let's Get Started

Grok is not the only good AI assistant, and this book will not pretend it is — Chapter 3 places it honestly beside the others. What Grok is, more than any of them, is *present* — an assistant that knows what happened this morning, draws on the live pulse of the world's largest real-time conversation, and folds the present into the questions you are allowed to ask. That advantage is real, and it comes bound to real risks, and using it well means holding both at once.

Turn the page. Let's start with what Grok is.

Part I: Meet Grok

Audience: everyone.

"Most assistants can tell you what was true. The interesting question is what's true right now — and whether your tool is even allowed to ask it."

— A working principle

Chapter 1: What Grok Is

I asked three different AI assistants the same question on a Tuesday morning: "What's the reaction so far to the announcement that came out an hour ago?" Two of them told me, politely, that they couldn't help — their knowledge had a cutoff months in the past, and the thing I was asking about hadn't happened yet as far as they knew. The third one answered. It pulled what people were actually saying right then, summarized the split between the people cheering and the people groaning, quoted a couple of the sharper takes, and noted that the official account hadn't yet responded. That third one was Grok.

Nothing about that is magic, and that's exactly the point worth sitting with. The other two assistants are not worse tools — for a great many tasks they are superb, and parts of this book will tell you honestly when to reach for them instead. But for the specific job of *knowing what is happening right now*, they were structurally unable to help, and Grok was built from the ground up to do precisely that. It is wired into a live stream of human conversation and current events, and it treats "what is true at this moment" as a question it can go and answer rather than one it has to apologize for.

That is the thing to hold onto as we start. This chapter is about what Grok is, what it isn't, and why a tool whose defining trait is *living in the present* turned out to be a genuinely different kind of assistant rather than just another chatbot with a different logo.

The One-Sentence Definition

Grok is xAI's AI assistant: a frontier conversational model — Grok 4.3 is the flagship as of this writing — that answers questions, writes, reasons, and codes, and, distinctively, reaches into real-time information from X and the live web to tell you what is true *now* rather than only what was true when it was trained.

A few words in that sentence carry most of the weight.

"Frontier conversational model" — Grok is not a toy or a novelty bolted onto a social network. It is a serious, top-tier model that competes directly with the strongest assistants from OpenAI, Anthropic, and Google on reasoning, coding, and writing. If you have dismissed Grok as "the edgy chatbot inside X," the first thing this book asks you to do is set that framing down. The tone story is real and we will give it an honest chapter (Chapter 24), but it is not the center of what Grok is.

"Real-time information from X and the live web" — this is the part the competitors do not match in the same structural way, and it is the reason Grok exists as a separate thing rather than a me-too. xAI was built by the company that owns X, and Grok has native, first-class access to the firehose of what people are posting and discussing as it happens. When you ask Grok about a breaking story, a live event, a just-released product, or the mood around something this afternoon, it can go look — at the actual posts, in real time — and tell you. That is a different capability from a model that was frozen at a training cutoff and can only reconstruct the past from memory.

"Tell you what is true now" — the output you get from Grok on a current-events question is not a recollection; it is a fresh read of live sources, ideally with the receipts. This changes what the tool is *for*. An assistant frozen at training time is a brilliant reference librarian who stopped reading the news on a particular date. Grok is a librarian who also has today's papers open on the desk.

"Sees, generates pictures and video" — Grok is multimodal in both directions. It can look at an image or a document you give it and reason about it (Chapter 15), and it can create images through Aurora (Chapter 12) and short video through Grok Imagine (Chapter 13). It is not a text-only engine with a search box.

What Grok Is Not

Five clarifications up front, because each is a misconception I have watched send people in the wrong direction.

Grok is not just "the AI inside X." It launched there and it is still deeply tied to X, but Grok today is reachable on its own website, in its own mobile apps, and through a developer API — not only as a feature of a social network. You do not need to live on X to use Grok, and treating it as merely an X feature undersells a standalone, top-tier assistant. Chapter 6 walks through every place you can reach it.

Grok is not defined by its tone. The most-repeated thing about Grok in the press is that it is "less filtered" or "has more personality" than its rivals. There is something real there, and Chapter 24 treats it seriously — including where that trait helps, where it has caused genuine trouble, and how to think about it at work. But a tool is not its reputation. Underneath the personality story is a capable, broadly useful assistant, and if you let the tone narrative dominate your picture of Grok you will misjudge what it is good for.

Grok is not a replacement for your judgment. Like every large language model, Grok can be confidently wrong, can misread a source, and can present a shaky inference with the same fluency as a solid fact. Real-time access makes some kinds of error *less* likely — it is harder to be out of date when you are reading live sources — but it introduces others, because live information is noisy, unverified, and sometimes flatly false. The live web includes rumor and propaganda alongside truth. Chapters 23 through 26 are the honest accounting of where Grok goes wrong and how to use it responsibly. The short version: real-time cuts both ways, and the section just below says exactly how.

Grok is not free in the way that matters most. There is a free tier, and it is genuinely useful for trying the tool. But the heaviest capabilities — the most capable model behavior, the deepest research mode, the highest usage ceilings — sit behind paid tiers, some of them tied to an X subscription and some sold directly by xAI. Chapter 5 maps exactly what each tier unlocks so you do not build a habit around a feature you cannot actually reach. I am flagging this now because nothing is more frustrating than reading about a workflow that turns out to be on the other side of a paywall.

Grok is not standing still, and neither are its numbers. The model name, the tier structure, the prices, the exact feature list — all of it moves on a scale of weeks to months. Grok 4.3 is the flagship as I write; it may not be by the time you read this. Where a detail is the kind that changes fast, I flag it and point you at xAI's current documentation rather than pretending the figure is permanent. The durable parts — what real-time AI is *for*, how to use it well, where to trust it and where not to — will hold.

Why "Real-Time" Is the Whole Game

It is tempting to evaluate an AI assistant by a single benchmark score or a single clever demo. Grok's real advantage is harder to put on a leaderboard and more important in practice: it collapses the gap between *when something happens* and *when your assistant knows about it.*

Every model frozen at a training cutoff has the same fundamental blind spot — the present. Ask it about this morning's news, the reaction to a product that shipped an hour ago, the score of a game still being played, the sentiment around a stock this afternoon, and the honest answer is some version of "I don't know; my information ends on a date in the past." You can paper over that with bolted-on web search, and the other assistants increasingly do. But Grok was *designed* around the present rather than retrofitted for it, and it has a source the others do not: native, real-time access to X, where an enormous amount of the world's reaction to events shows up first, in human words, as it happens.

That changes the kinds of questions worth asking. With a frozen assistant you learn not to bother asking about anything recent, the way you learn not to ask a library book about tomorrow's weather. With Grok you can ask "what's the reaction to X," "what are people saying about Y right now," "did Z actually ship yet," "what's the live sentiment on this" — and get a real answer drawn from real, current sources. The capability is not just "search is available." It is that *the present is a first-class thing the tool reaches for by default*, which over time changes what you bring to it.

Watch how that plays out in an ordinary afternoon and the difference stops being abstract. A piece of news flickers across your feed — a company you follow is rumored to have shipped something, or pulled something, or said something. With a frozen assistant your only move is to go do the legwork yourself: open a browser, run searches, skim a few articles, try to gauge whether the thing is real and how big a deal it is. With Grok the question itself becomes the task: "Did this actually happen, and what's the reaction?" It goes and reads the live conversation, tells you whether the rumor is confirmed or still just a rumor, and gives you the texture of how people are taking it — measured, alarmed, indifferent — with sources you can open. The thing you used to do *around* the assistant becomes the thing you ask it to do. That is a quiet but real change in workflow, and it accrues: over weeks, you stop pre-filtering your questions for "is this too recent to ask," and you simply ask.

The flip side deserves equal billing, because the same property that makes Grok useful here makes it newly fallible. Live sources are unverified by definition — the post Grok reads thirty seconds after an event may be wrong, may be a guess, may be deliberate misinformation riding the moment. A frozen model's errors tend to be errors of staleness, which you can anticipate. A real-time model adds errors of *velocity*: it can faithfully report what is being said while what is being said is itself false. This is not a reason to avoid the capability; it is a reason to use it the way a good journalist uses a live wire — as the fastest possible read on what is being claimed, held a little loosely until it firms up. The book returns to this with real care in Chapters 23 through 26, but it belongs in the first picture of the tool: the present is a superpower and a fresh source of error in the same motion.

That is why the real-time and search material gets its own Part (Part III) and why this chapter keeps returning to it. It is not a feature brag. It is the mechanism that makes Grok a different tool rather than a differently branded one.

What Grok Actually Does

Set the real-time story aside for a moment and look at the full capability surface. Given a request, Grok can:

Hold a real conversation and reason carefully. It answers questions, explains, argues, drafts, and — when the problem is hard — switches into an explicit Think mode that reasons step by step before answering (Chapter 8). It is strong on the things frontier models are now strong on: analysis, writing, math, code.

Search the live web and X in depth. Beyond a quick lookup, its DeepSearch and DeeperSearch modes run extended, multi-source research over current information and hand you back a synthesized, sourced answer (Chapter 9). This is the real-time advantage turned into a workflow.

See and read. Give it a photo, a screenshot, a chart, or a document, and it can describe, extract, and reason about what is in it (Chapter 15).

Create images and video. Aurora generates images from a text description (Chapter 12); Grok Imagine produces short video (Chapter 13). Creation sits beside comprehension.

Talk with you out loud. A voice mode lets you speak to Grok and hear it answer, for hands-free and conversational use (Chapter 14).

Be built upon. A developer API exposes the models, real-time search, and tool use to your own software, so Grok's capabilities can live inside applications you build rather than only inside the chat box (Part VI).

It is worth seeing how these pieces combine, because the interesting work is rarely a single capability used alone. Picture a small, realistic task: you are preparing for a meeting about a competitor's new product. A single request can have Grok run a DeepSearch across the last few days of coverage and live reaction, read the screenshot you paste of the competitor's pricing page, reason through how their pricing compares to yours, and draft a one-page brief you can hand to your team — research, vision, analysis, and writing chained into one outcome. None of those steps is exotic on its own; the value is that they compose, and that the research leg can reach into this week rather than stopping at a training cutoff. That composability, with the present available as one of the ingredients, is closer to what daily use of Grok actually feels like than any single-feature demo.

What it produces, across all of these, is meant to be checked and used — not swallowed whole. Especially when the answer rests on live sources, the right posture is to glance at where the claim came from. Grok is at its best when you treat it as a fast, current, capable colleague whose work you still read.

Who's Using Grok

Roughly three overlapping groups, in my experience.

Individuals who want a capable everyday assistant that also happens to know what is going on right now — for keeping up with a fast-moving topic, checking live reactions, research that touches recent events, and the ordinary write-summarize-explain-code work every assistant does.

Professionals and teams who lean on the real-time strength for a specific job: monitoring sentiment and breaking developments, research that has to include the last twenty-four hours, analysis where being current is the whole value. For these readers the live edge is not a novelty; it is the reason to choose Grok for the task.

Developers and builders wiring Grok's API — and especially its live-search tools — into their own products, so that applications can answer questions about the present, not just the past (Part VI is written with this group in mind).

This book serves all three, and the parts are arranged so you can read the one that is yours and skim the rest.

The Through-Line

If there is a single idea to carry out of this chapter, it is the one the morning-of-the-announcement story opened with: most AI assistants live in the past, and Grok is built to live in the present. Everything distinctive about it — DeepSearch, the X integration, the way it answers "what's happening right now" instead of apologizing for not knowing — flows from that one design choice. The tone, the personality, the controversy: real, and honestly covered later, but secondary. The center is real-time.

The rest of Part I fills in the picture — where xAI and Grok came from, how Grok stacks up against the assistants you may already use, and what the Grok 4.x models under the hood actually are. For now, the one-sentence version is enough to build on: Grok is the frontier assistant that knows what happened this morning, and using it well starts with knowing when "this morning" is the question that matters.

Chapter 2: The xAI Story

Most AI assistants are made by companies whose business is AI. Grok is made by a company that also owns the largest real-time conversation platform in the world, and that single fact explains more about Grok than any benchmark does. To understand why Grok is the assistant that lives in the present — and why it has the particular strengths and the particular controversies it has — you have to understand where it came from. This chapter tells that story, because the gap between Grok and its rivals is, more than anything, a gap in origin.

A Company Built to Have an Opinion

xAI was founded in 2023 by Elon Musk, who had been an early backer of OpenAI before parting ways with it years earlier. The founding pitch was deliberately pointed: that AI was being built in a way Musk considered too cautious, too filtered, and too willing to refuse, and that the world needed an alternative that was more willing to engage, more "maximally truth-seeking," and less hedged. Whatever you make of that framing — and reasonable people land all over it — it is the founding DNA, and it shows up directly in how Grok behaves and in the debates that follow it around. Grok's personality is not an accident of training; it is a stated design goal traceable to the day the company started.

The second founding fact matters just as much: xAI was created inside Musk's orbit alongside X, the platform formerly known as Twitter, which he owns. That proximity is the source of Grok's signature capability. An AI company without a social network has to negotiate for access to real-time human conversation, scraping or licensing it at arm's length. xAI had a live firehose of it in-house from day one. The real-time advantage that runs through this whole book is, at bottom, a corporate-structure advantage: the assistant and the conversation platform share an owner.

It is worth dwelling on how unusual that arrangement is, because the rest of the field has been straining against the exact constraint xAI was handed for free. Consider the position of a competitor. OpenAI, Anthropic, and Google all want their assistants to know about the present, and all three have built or bought their way to some version of web search. But none of them owns a platform where hundreds of millions of people post their reactions to events in the moment those events happen. When a competitor wants to know "what is the world saying about this right now," it has to send a crawler out to the open web and hope the relevant conversation has been indexed somewhere it can reach — and the place that conversation most often happens first is precisely the platform it does not own. X has at various points restricted or charged heavily for firehose-level access to its data, which means the very signal everyone wants is the one signal xAI gets at cost and everyone else gets late, partial, or not at all. The moat is not a clever algorithm. It is a deed of ownership.

That ownership also changes what "real-time" can mean in practice. A search index is a snapshot of the web taken some minutes or hours ago and refreshed on a schedule. A native tie to the posting stream is closer to a live tap. When a major story breaks — an earnings surprise, a sports result, a sudden resignation — the gap between "it happened" and "people are posting about it" is often seconds, and the gap between "people are posting" and "a third-party crawler has indexed those posts" can be much longer. Grok sits on the near side of that second gap by construction. This is why, throughout the book, I treat real-time not as one feature among many but as the organizing fact of the whole tool. It is the one thing the corporate structure makes structurally easier for Grok than for anyone else.

From Announcement to Open Weights

Grok went public late in 2023, launched first to paying X subscribers as a feature of the platform — which is why so many people still think of it as "the X chatbot." The earliest version was a capable-but-not-frontier model with a distinct, jokey persona and a built-in connection to X's live posts. It was rough, it was opinionated, and it was clearly version one of something, not a finished product.

Then xAI did something that surprised a lot of people: in early 2024 it open-sourced the weights of that first model, Grok-1, releasing them publicly under a permissive license. The model was enormous — a mixture-of-experts design with hundreds of billions of parameters — and far too large for most people to run, so the practical impact was limited. But the gesture was real and worth remembering, because it sits in tension with the more closed posture the company took as the models got more valuable. xAI has, at various points, positioned itself as friendlier to openness than its rivals; the reality has been mixed, the way it usually is across this industry. The honest summary is that Grok-1's weights are out there, and the frontier models that followed are not.

It helps to understand why "open-sourcing the weights" of a hundreds-of-billions-parameter model was more symbol than gift. Releasing weights is not like releasing a small program you can run on a laptop. To actually load and run a model of that size you need a cluster of high-end accelerators worth far more than a car, plus the engineering to shard the model across them and keep it fed. So when xAI published Grok-1, the set of people who could meaningfully use the release was small — a handful of well-funded labs and the most serious independent researchers. What the release did accomplish was reputational and directional: it planted a flag that said "we are on the open side of the open-versus-closed argument that was consuming the field in 2024," at a moment when Musk was publicly and litigiously criticizing OpenAI for having drifted from its own open founding ideals. Read in that light, Grok-1's release was as much a move in an argument as a contribution to the commons.

What followed is the pattern you should expect rather than be surprised by. As each new generation got more capable — and therefore more commercially valuable and more expensive to train — the weights stopped going out the door. This is not hypocrisy unique to xAI; it is the gravity that has pulled almost every lab the same direction. The frontier costs hundreds of millions of dollars to reach, and giving away the result for free is a hard business to run. The durable lesson for you as a user is practical: do not choose Grok expecting to download and self-host the model that actually powers the product. You will reach the frontier Grok through xAI's apps and its API, the same way you reach the frontier models of the other labs. The openness in Grok's story is real history, not a present-day feature you can lean on.

The Climb to the Frontier

What happened next was a fast, well-funded climb from novelty to genuine contender. The arc is easy to lose in version numbers, so here is the shape of it rather than a changelog.

The models got serious. Across successive generations — Grok-1 to 1.5 to 2 to 3 to 4 and into the 4.x line — Grok moved from "fun but not frontier" to competing head-to-head with the best models from OpenAI, Anthropic, and Google on reasoning, math, coding, and knowledge. By the Grok-3 and Grok-4 era, dismissing Grok as a lightweight was simply out of date; it posts top-tier results on the benchmarks the field watches, and the day-to-day experience backs that up. Chapter 4 covers the current generation in detail.

The capabilities broadened. Grok grew from a text chatbot into a multimodal system. Image understanding arrived, so it could see. Image *generation* arrived through xAI's own Aurora model rather than a licensed third-party engine. Then short-form video generation arrived as Grok Imagine. A voice mode let you talk to it. The product went from "a chat box" to "a creative and conversational suite," and Part IV of this book is largely about that breadth.

Real-time search became the headline. The capability that always distinguished Grok — its tie to X — matured into a deliberate research product. DeepSearch and DeeperSearch turned "Grok can see X posts" into "Grok can run an extended, multi-source investigation across live information and hand you a sourced synthesis." This is the through-line of the whole company made into a feature, and Part III is devoted to it.

It is easy to read that arc as a tidy march of progress, so it is worth being honest about what "competing head-to-head" did and did not mean at each step. In the early days the gap was obvious to anyone who used the tools side by side: the first Grok was fun and current but visibly behind on the hard stuff — give it a genuinely difficult reasoning problem or a non-trivial coding task and it would stumble where a leading model would not. That gap closed in stages rather than overnight. The Grok-2 era narrowed it; by Grok-3 and Grok-4 the day-to-day experience on ordinary work became hard to distinguish from the best of the field, and on the benchmarks the labs watch most closely Grok began posting numbers at or near the top. The point of the story is not that Grok was always great — it was not — but that the trajectory was unusually steep, and that the version you can use today is a fundamentally more serious instrument than the one that earned the early "X chatbot" reputation. If your mental image of Grok was formed in 2023 or 2024, it is simply out of date, and Chapter 4 is where the current generation gets its proper accounting.

Colossus: Building the Engine

One part of the xAI story is genuinely unusual and worth knowing, because it explains how a company founded in 2023 caught up to rivals with years of head start. xAI built, at remarkable speed, one of the largest AI training supercomputers in the world — a cluster nicknamed Colossus, stood up in Memphis, packed with a vast number of high-end GPUs and expanded aggressively after that. The build-out was fast enough to raise eyebrows across the industry, and it drew real criticism on environmental and local-impact grounds that are part of the honest record. But the strategic point is simple: training frontier models takes enormous compute, and xAI chose to own a colossal amount of it rather than rent its way to the frontier. The rapid model progress in the last few sections is downstream of that bet on raw infrastructure.

The reason this matters to you, a user, rather than only to industry watchers, is that compute is the rate limiter on how good a model can get and how fast. Training a frontier model is, simplifying only a little, a matter of running an enormous amount of arithmetic over an enormous amount of data, and the speed at which you can run that arithmetic is set by how many high-end chips you can point at the problem at once. A lab that has to rent its chips from a cloud provider waits in the same queue as everyone else, pays a margin, and is capped by whatever capacity it can secure. A lab that owns a dedicated cluster can iterate faster, train larger, and retrain more often. When you notice that Grok ships new model revisions on a cadence of weeks to months — fast enough that this book has to flag version numbers as perishable — that velocity is not marketing energy. It is the visible output of owning the engine room. The speed of the product follows from the speed of the training, and the speed of the training follows from Colossus.

There is an honest cost to the speed, and a guide should name it. Standing up that much computing capacity that quickly drew sustained criticism over its power and water demands and its impact on the Memphis community where it was built, including reporting and complaints about emissions from on-site generators. None of that changes how the tool behaves on your screen, and I am not going to pretend to adjudicate the disputes here. But the same aggressive, move-fast posture that produced a frontier model in record time is the posture that produced those frictions, and a clear-eyed reader holds the two together rather than admiring one and ignoring the other. The infrastructure story, like the personality story, is a place where Grok's strengths and its controversies grow from the same root.

The Controversies Are Part of the Story

A hagiography would skip this section. A useful guide cannot, because Grok's rough edges are not incidental — they flow from the same founding choices that give it its strengths, and you cannot understand the tool without them.

Grok has, on more than one occasion, generated outputs that caused real public controversy: offensive content, politically charged answers, episodes where the "less filtered" posture produced material that xAI itself had to walk back and correct. Some of these were tied to deliberate changes in how the model was instructed to behave; some were the model behaving as a less-restricted model sometimes will. The company's relationship to content moderation, to its founder's own very public political profile, and to the question of what "truth-seeking" means in practice has been a recurring source of friction and news.

The mechanism behind these episodes is worth understanding plainly, because it demystifies what can look like random misbehavior. A modern assistant's tone and boundaries are not fixed properties of the underlying model so much as the product of two adjustable layers on top of it: the training that shapes its default disposition, and a system prompt — a standing set of instructions the model receives before your conversation begins — that tells it how to carry itself. Tug either layer toward "be less hedged, engage more, refuse less," and you move the whole distribution of its outputs in that direction, including the tail of the distribution where the genuinely bad outputs live. Several of Grok's public stumbles trace directly to changes at exactly those layers; in at least one well-publicized instance the company attributed problematic behavior to an unauthorized or ill-considered modification of the system instructions and rolled it back. The lesson is not that Grok is uniquely unsafe — every lab tunes these dials — but that the dials are real, their settings have consequences, and a tool tuned deliberately toward directness will, predictably and occasionally, overshoot.

I am not raising this to editorialize, and Chapter 24 treats the "less filtered" question carefully and without either cheerleading or panic. I raise it here because it belongs in the origin story. The same DNA that makes Grok willing to engage where other assistants refuse, and that gives it a livelier voice, is the DNA that has periodically produced trouble. Strength and liability share a root. A clear-eyed user holds both facts at once: Grok's directness is genuinely useful for many tasks, and its looser guardrails are a real consideration — especially at work, where Chapter 26 picks the question back up.

Why the Origin Explains the Product

Step back and the whole arc resolves into a few through-lines that you will feel every time you use the tool.

Grok lives in the present **because** it was born inside a real-time conversation platform. The live-knowledge advantage is corporate structure made into capability.

Grok has a distinct voice and looser guardrails **because** its founding thesis was that AI had become too cautious. The personality and the controversies are two faces of one design choice.

Grok reached the frontier fast **because** xAI poured money and enormous self-owned compute into catching up, building Colossus rather than waiting in line for someone else's hardware.

And it grew into a full creative and conversational suite — vision, Aurora, Grok Imagine, voice — for the plain reason that the company kept widening the product rather than settling for a text box.

None of that is marketing. It is just what you get when you trace the tool back to the company that made it: an assistant built to live in the present, to say what it thinks, and to reach the frontier in a hurry — wearing every one of those choices in public. Which leaves the question any sensible person asks before adopting anything new: granting that Grok is real and capable, is it actually better than what I already use? Not always, and not for everything — and knowing exactly when it is, is most of the skill.

Chapter 3: Grok vs. ChatGPT, Claude, and Gemini

The honest answer to "should I use Grok instead of ChatGPT, Claude, or Gemini?" is "instead of, rarely; alongside, often." This book is about Grok and it will make Grok's case, but it is a guide, not an advertisement, and a guide that pretended Grok was the right tool for every job would be lying to you. The four leading assistants overlap enormously — any of them can write your email, summarize your document, explain a concept, or draft your code, and for most everyday tasks the differences between them are smaller than the marketing on any side suggests. What follows is a clear-eyed account of where Grok genuinely pulls ahead, where it doesn't, and how a sensible person decides which tool to reach for.

Start With What's the Same

Before the differences, the unglamorous truth: at the frontier, these models are more alike than different. ChatGPT (OpenAI), Claude (Anthropic), Gemini (Google), and Grok (xAI) are all top-tier conversational models that handle the bulk of ordinary work — writing, summarizing, explaining, analyzing, coding, answering questions — at a quality that would have seemed impossible a few years ago and that, on a blind test of a typical task, most people could not reliably tell apart. If you already have a paid assistant you are happy with, Grok is not going to feel like a different species. It will feel like a strong peer with a couple of distinctive talents.

That framing matters because it sets the right expectation. The question is almost never "which single assistant is best at everything" — none is — but "which one is best for *this* task, and is its particular edge worth a second subscription." With that lens, here is where Grok's edge actually lives.

A quick worked example makes the point concrete. Suppose you need to draft a polite email declining a meeting, summarize a six-page PDF into three bullet points, and explain how a Roth conversion works. Run those three tasks through all four assistants and the honest result is that you would struggle to rank the outputs — they would all be competent, the differences would come down to small matters of phrasing and structure, and your personal preference would matter more than any objective gap. Now change the task to "what did the Federal Reserve actually say at the meeting that ended an hour ago, and how are markets reacting." Suddenly the field splits. The assistants that are frozen at a training cutoff either decline or, worse, reconstruct a plausible-sounding answer from old information; the ones with bolted-on search do better but may surface a stale article; and the one built around the live stream of reaction is in its element. The lesson is not that Grok is better — it is that the differences only become decisive on a specific slice of tasks, and the skill is recognizing when you are on that slice.

Where Grok Genuinely Pulls Ahead

Real-time knowledge — this is the big one. Grok's defining advantage is live information, and especially its native, first-class access to X. When the task depends on *now* — breaking news, the reaction to something that just happened, live sentiment around a topic, whether a thing has actually shipped yet, what people are saying this afternoon — Grok is structurally better positioned than the others. The competitors have all added web search, and it works, but Grok was built around the present rather than retrofitted for it, and it draws on the firehose of human reaction that lives on X. For "what's happening right now" questions, Grok is usually the one I reach for. Part III is the whole story.

X-native understanding. Beyond raw freshness, Grok understands X itself — its accounts, its threads, its culture, the shape of a conversation happening on the platform — in a way an outside model peering in through a search API does not. If your question is *about* X, or about a discourse that is unfolding on X, that native fluency is a real edge.

The difference shows up in the texture of the answer, not just its recency. Ask an outside model "what are people saying about the new phone," and at best it returns a digest of whatever articles a crawler happened to index — secondhand, already filtered through journalists, and skewed toward whatever ranks in search. Ask Grok the same thing and it can read the primary reaction: the actual posts, the ratio of enthusiasm to complaint, which specific criticisms are gaining traction, which accounts are driving the conversation, whether the loudest take is coming from a handful of large accounts or from a genuinely broad base. That is a qualitatively different kind of knowing. It is the difference between a poll summary and standing in the room. The same edge applies when the subject *is* the platform — tracing how a particular thread spread, understanding a piece of platform-native slang, or reading the shape of a pile-on as it forms. None of this means the raw posts are true; the live web is full of noise and bad actors, a limit Chapter 10 and Part VII take seriously. It means Grok can show you the unfiltered conversation rather than a sanitized report of it, and for many real questions that primary view is exactly what you want.

Directness and a willing voice. Grok is, by design, more willing to engage with edgy, contentious, or blunt requests and to answer in a livelier register than its more cautious rivals. For some users and some tasks — wanting a straight take rather than a hedged one, a bit of personality rather than corporate neutrality — that is genuinely the appeal. It is also, as Chapters 24 and 26 cover honestly, a double-edged trait that needs judgment, especially at work. But for the person who finds the other assistants over-cautious, this is a real and intentional difference, not a bug.

Where Grok Doesn't Lead

A guide you can trust has to say this part too.

Ecosystem and integration depth. ChatGPT and Gemini are embedded in vast, mature ecosystems — ChatGPT in OpenAI's enormous developer and product world, Gemini woven directly into Google Workspace, Android, and Search. If your life runs in Gmail, Docs, and the rest of Google, Gemini's integration is a convenience Grok cannot match. ChatGPT's sheer breadth of features, plugins, and third-party tooling is similarly hard to equal. Grok is catching up on surfaces and integrations, but "it's already inside the tools I use all day" is a real advantage the incumbents hold.

Some specialized strengths. Each rival has tasks it is particularly loved for. Claude has a strong reputation for careful long-form writing, nuanced reasoning, and a measured, low-drama voice that many professionals prefer for serious work; it is also widely favored for certain coding workflows. ChatGPT has the deepest bench of built-in tools and the largest community of shared techniques. Gemini brings very large context handling and tight Google-data integration. Grok holds its own across all of these, and leads on real-time — but "best in class" for a given specialized task is often one of the others, and pretending otherwise would not help you.

It is worth being concrete about what these preferences feel like in practice, because they are easy to dismiss as fanboyism and they are usually more grounded than that. Take long-form writing. People who draft serious documents — a board memo, a legal brief, a chapter — often reach for the assistant whose default voice needs the least correction, because the cost of a tool is not just the quality of its first draft but the amount of editing that draft demands. A model with a steady, plain register saves you the work of stripping out cheerfulness and filler; that is a real, repeated time saving, and it is why a measured voice earns loyalty. Take coding. The thing developers actually value is not raw cleverness on a one-off puzzle but reliability across a long session — does the assistant keep track of the codebase you handed it, follow your conventions, and avoid quietly breaking something three files away — and reputations there are built on hundreds of hours, not benchmarks. Take Google integration. If your calendar, your mail, and your documents already live in Workspace, an assistant that can act on them in place removes a dozen copy-paste steps a day, and that convenience compounds. None of these are knockout advantages, and Grok is competitive enough on all of them that you will rarely feel actively held back. But when someone tells you they prefer one of the incumbents for a specific kind of work, they are usually pointing at a genuine fit between the tool's character and the task, and the grown-up move is to respect that rather than argue them out of it.

The maturity and trust story. The incumbents have longer track records, more settled enterprise and compliance stories, and — relevant to Grok specifically — fewer public controversies over their outputs. For risk-averse organizations, that track record is itself a feature, and Grok's looser-guardrails reputation (Chapter 24) is something a cautious buyer weighs. This is not a capability gap; it is a trust-and-governance gap, and it is real.

How to Actually Choose

Forget "which is best" and ask "which for this." A few practical rules of thumb, all of which you should treat as starting points rather than laws:

Reach for Grok when the present is the point. Anything that hinges on current events, live reaction, breaking developments, or what is being said on X right now — Grok first. This is its home turf and the clearest reason to have it.

Reach for Grok when you want a direct, less-hedged take and you have the judgment to handle one — and you are not in a context (regulated, public-facing, sensitive) where the looser register is a liability.

Reach for the incumbent that already lives in your workflow for ordinary daily tasks. If you are deep in Google Workspace, Gemini's integration is hard to beat for in-context work; if your team standardized on ChatGPT or Claude, the friction of switching for routine tasks usually is not worth it.

Reach for the incumbent when the answer must not move. This is the inverse of the real-time rule and it is just as important. For a question whose correct answer is settled and stable — a point of established law, a mathematical method, the syntax of a programming language, the plot of a published book — pulling in live web reaction adds noise rather than signal, and you are better served by a model answering from its trained knowledge. Real-time is a superpower for moving targets and a liability for fixed ones; matching the tool to whether the answer can change is most of the skill.

Use more than one. The quietly correct answer for a lot of serious users is not to pick a single assistant but to keep two — a primary all-rounder and Grok for the real-time work it does best. The assistants are cheap enough relative to the value of being right that running a primary plus Grok-for-now is a reasonable setup, not an indulgence. Much of this book assumes you may be doing exactly that.

If a two-tool setup sounds like overhead, picture how it actually plays out across a working day. You live in your primary for the bulk of the work — the drafting, the summarizing, the explaining, the coding — because switching tools for routine tasks is friction with no payoff. Then a question arrives that turns on the present: a client asks whether a competitor has really launched the thing the rumor says they launched; a story is breaking in your industry and you need the shape of it before a call in twenty minutes; you want to know whether the outage everyone is complaining about is widespread or just you. Those are the moments you reach for Grok, and the habit that serves you is a simple mental trigger: *if the answer could have changed since this morning, ask the assistant that reads this morning*. For everything else, stay where you are. A useful sharpening of this is to also let the two tools check each other on a claim that matters — have one draft an answer and the other critique it — which catches a surprising number of confident errors precisely because the models fail in different ways. The point of keeping two is not to use both for everything; it is to have the right one ready when the task quietly demands it.

The Comparison That Matters Is Yours

Benchmark leaderboards move every few weeks, and any specific "Grok beats X on benchmark Y" claim I put here would be out of date within a quarter — so I am deliberately not anchoring this chapter to scores. The comparison that actually matters is not on a leaderboard; it is the one you run on your own real tasks. Take a week of your genuine work — the questions you actually ask, the documents you actually summarize, the things you actually need to know — and run a handful of them through Grok and through whatever you use now. You will learn more from that than from any benchmark table, and you will quickly feel where Grok's real-time edge earns its place in your kit and where your incumbent still wins.

What you should take from this chapter is the lay of the land. Grok is a genuine frontier assistant — first among equals on real-time and X-native work, fully competitive everywhere else, and behind only on ecosystem depth and settled trust. Best understood as a powerful specialist-generalist you add to your kit, not a wholesale replacement you switch to. Hold that, and you will deploy it where it shines instead of forcing it where it merely ties.

Chapter 4: The Grok 4.x Model Generation

Everything Grok does — the conversation, the reasoning, the real-time research, the code, the analysis — is the work of a model, and the model is where the capability ultimately lives. You can use Grok well without thinking about the model much, the same way you can drive without knowing the engine. But a little understanding of what is under the hood pays off, because it tells you what to expect, what the dials do, and which numbers to care about. This chapter is that understanding, and it closes Part I. One caution before we start: this is the most perishable chapter in the orientation, because model names and numbers change faster than anything else in this book. I will tell you the durable shape and flag the specifics.

Grok 4.3 Is the Flagship — As of This Writing

The flagship model as I write is **Grok 4.3**, the leading member of the Grok 4.x generation. That qualifier — *as of this writing* — is load-bearing. By the time you read this there may be a 4.4, a 4.5, or a Grok 5; xAI ships new and updated models on a cadence measured in weeks to months, and the name on the current flagship is exactly the kind of detail that drifts. So rather than fixate on the version string, learn the generation's character, which changes more slowly than its number.

The Grok 4.x generation is a frontier-class line: top-tier on the reasoning, math, coding, and knowledge tasks the field uses to rank models, and the basis for Grok competing head-to-head with the strongest assistants from the other labs (Chapter 3). When you talk to Grok in its default, capable mode, this is what is answering. **Confirm the current model name and the present lineup against xAI's own documentation** before you rely on a specific version — the live page is always the authority, and I would rather you check than trust a number that has aged.

It helps to know why these numbers move as fast as they do, because the churn is not chaos — it follows a rhythm you can learn to read. A model generation is not one event but a sequence: an initial release, then a stream of point revisions that fix weaknesses, extend context, speed up inference, or sharpen a particular skill, often shipped without much fanfare. The version string you see — 4.2, 4.3, and so on — is xAI's way of marking those increments. For you this has a concrete consequence. The "4" tells you the generation, which is the part that holds steady for a good while and that determines the broad character of what you are talking to. The digit after the dot tells you the increment, which is the part that drifts under your feet. So when you read "Grok 4.3 is the flagship," carry forward the *4* as durable and treat the *.3* as a snapshot. If you sit down to use Grok and the interface says you are on something else in the 4.x line, you are not on the wrong tool or reading an obsolete book; you are simply a few increments along from where this page was written.

A Family, Not a Single Model

"Grok 4.x" is better understood as a family than a single thing, and the members trade off along the lines every model family does:

Capability versus cost. More capable behavior costs more per unit of work, whether you are paying in subscription usage or in API tokens. The flagship is the strongest and the priciest; lighter or faster variants cost less and are plenty for routine work.

Capability versus speed. The most capable behavior — especially deep reasoning — takes longer to produce. A quick answer and a hard, carefully reasoned one are not the same latency, and that is a feature, not a flaw: you spend the time where the difficulty is.

Generalist versus specialized. xAI has, at points, fielded variants tuned for particular jobs — a faster everyday model, a heavier reasoning model, models oriented toward coding or toward specific surfaces. The exact roster of named variants shifts with each release, so treat the *idea* of a family as durable and the *specific members* as something to check against current docs.

To make the trade-offs less abstract, here is how the same family looks from three different seats. A casual user on a phone never picks a variant at all; they get whatever the app defaults to, which is tuned to feel fast and capable for ordinary questions, and that is exactly the right experience for someone who just wants a good answer without thinking about it. A developer building a product, by contrast, picks the variant deliberately for each call: the lighter, cheaper model for the high-volume, low-stakes work — classifying incoming messages, summarizing routine text, the thousand small calls a real application makes — and the flagship reserved for the requests where the quality difference actually shows. And a power user running heavy research sessions might lean on the most capable reasoning variant precisely because the time and cost are worth it when the question is hard and being right matters. Same family, three sensible defaults, each matched to what the seat actually needs.

The mistake worth avoiding in all three seats is reflexively reaching for the most powerful option on every task. The flagship is not strictly "better" the way a faster car is faster; on a simple lookup it is slower and more expensive for an answer no human could distinguish from the lighter model's. Capability you do not need is not a benefit, it is a tax — paid in latency for the impatient user and in dollars for the developer running at scale. Right-sizing the model to the task is a real skill, and it is the same skill whether you are choosing a mode in the app or a model name in an API call.

For most users most of the time, the right posture is the simple one: **stay on the capable default and let Grok do its thing.** The family exists so that heavy users, developers, and cost-sensitive workloads can pick a point on the trade-off curve — but you do not have to think about it to use Grok well.

The Reasoning Dimension: Think Mode

The single most useful thing to understand about this generation is that it can reason in two registers. For an ordinary question, Grok answers directly and fast. For a hard one, it can switch into an explicit, deliberate reasoning mode — **Think mode** — where it works through the problem step by step before committing to an answer, planning and checking rather than replying off the top of its head.

What is actually happening when a model "thinks" is worth a plain-language picture, because it demystifies both the benefit and the cost. In its fast mode, the model produces its answer in more or less one pass — it reads your question and generates a reply, the way you might blurt out the first thing that comes to mind. In Think mode, it instead generates an extended internal chain of working before it writes the final answer: laying out the steps, trying an approach, noticing a problem, backing up, checking its arithmetic, considering a case it missed. That visible-or-hidden scratch work is why the answer is better on hard problems — the model has, in effect, given itself room to catch its own mistakes before committing — and it is also exactly why the answer takes longer and, through the API, costs more, because all that intermediate reasoning is real computation that has to be produced. You are quite literally buying correctness with time and tokens.

This matters because it is a dial you will actually touch. Hard logic puzzles, multi-step math, careful analysis, tricky code, anything where a fast guess is likely to be a wrong guess — these are where Think mode earns its keep, trading time for correctness. A worked feel for it: ask "what's the capital of Australia" and Think mode is pure waste, because the fast answer is already right and waiting adds nothing. Ask "here are four constraints on who can sit where at a dinner table — find a valid arrangement, or prove none exists" and the fast mode will often produce a confident, plausible, and wrong answer, while Think mode will actually grind through the cases and either find the seating or show you why it cannot. The tell that you are in Think-mode territory is usually that you can imagine yourself needing scratch paper. Routine questions do not need it; reaching for deep reasoning on a simple lookup just makes you wait. The heuristic is the one that governs all of frontier AI right now: **think harder for hard things, answer fast for easy ones.** Chapter 8 is devoted to Think mode and reasoning in practice; for now, just hold that the capability exists and is one of the defining features of the 4.x line.

Context: How Much Grok Can Hold at Once

A model's *context window* is how much text it can keep in mind in a single conversation — your prompt, the documents you have given it, and everything said so far, all counted together. When you exceed it, the earliest material falls out of the model's view, which is why a very long session can seem to "forget" how it began.

The Grok 4.x generation offers a **large context window — on the order of one million tokens** at the high end, which is a great deal of room: long documents, large codebases, extended research sessions, and lengthy back-and-forth all fit comfortably where they would have overflowed an earlier model. A million tokens is roughly several long books' worth of text held in mind at once. The practical effect is that you rarely have to think about the limit for ordinary work — you can paste the whole report, share the long thread, hand over the big file, and Grok keeps it all in view.

There is a subtlety here that separates people who use long context well from people who get burned by it, and it is worth stating plainly because the headline number invites overconfidence. A large window means the material *fits*; it does not guarantee the model attends to every part of it equally. In practice, models tend to track the beginning and the end of a long context more reliably than the vast middle — a tendency the field has discussed under the label "lost in the middle." So if you paste a hundred-page contract and ask about a clause buried in the middle, the model can in principle see it but may not weight it the way you would. The practical discipline that follows is simple and pays off: when something specific matters, do not rely on the model to find the needle in the haystack you handed it — quote or point to the relevant passage, or ask a pointed question about it, rather than trusting that "it's all in there" means "it's all equally in mind." Large context removes the old chore of trimming material to fit. It does not remove the judgment of telling the model what to look at.

Treat the exact figure as a check-the-docs number — context sizes change, and they sometimes differ by surface and by tier — but the durable point holds: this generation has room to spare for the kind of long-context work that used to require careful trimming.

The Numbers Worth Knowing — and the Flag on Them

Most of this book keeps dollar figures out of Part I deliberately, because consumer pricing and tier structure are volatile enough to deserve their own treatment (Chapter 5) and their own eyes-on confirmation. But two numbers belong here because they describe the model itself, and developers in particular will want them early.

For programmatic use through the API, the flagship's token pricing as of this writing is in the neighborhood of **$1.25 per million input tokens and $2.50 per million output tokens** — input being what you send the model, output being what it generates back. That ratio, output costing more than input, is standard across the industry, and these rates put the flagship in the competitive range for a frontier model. **Verify both figures against xAI's current API pricing before you budget against them** — API rates move, they vary by model within the family, and additional charges (for real-time search calls and other tools) can apply on top, as Chapter 22 details. The number here is a reference point, not a quote.

I am giving you these two and stopping there. The consumer subscription prices — what the various paid tiers cost a person rather than a program — are genuinely fluid right now, and I am not going to pin figures I cannot stand behind. Chapter 5 handles the access landscape properly, with the same honesty about what is durable structure and what is a number to confirm.

What to Carry Forward

You do not need to memorize a version string or a token count to use Grok well. What is worth carrying out of this chapter is the shape of the thing:

A **frontier-class family** — Grok 4.x, with 4.3 the flagship — strong across reasoning, coding, and knowledge, and the engine behind everything in the rest of this book.

A **reasoning dial** — direct answers for easy work, Think mode for hard work — that is the most useful control you will actually reach for, and the subject of Chapter 8.

A **large context window**, around a million tokens at the top end, big enough that long documents and long sessions just fit.

A set of **numbers that are real but perishable** — the API rates above, and everything in Chapter 5 — that you confirm against xAI's live documentation rather than trusting on a printed page.

That is the engine — and with it, Part I has done its job. You now know what Grok is, where it came from, how it stands beside its rivals, and what powers it. What you do not yet know is what it costs to get at, and that is where Part II begins: access, tiers, and the surfaces Grok lives on, then your first real conversations.

Part II: Getting Started

Audience: everyone. The setup is genuinely easy — but the access tiers matter, so read Chapter 5 before you spend anything.

"Don't decide whether the tool is for you by reading about it. Decide by asking it something only a tool that knows today could answer."

— A working principle

Chapter 5: Access and Tiers: Free, X Premium+, SuperGrok, and Heavy

Before you spend a cent, spend ten minutes understanding what you are buying, because Grok's capabilities are not uniformly available and the most-talked-about ones sit behind paid tiers. There is nothing more deflating than reading about a feature for three chapters and then discovering it is on the other side of a paywall you are not behind. So this chapter front-loads the access landscape. After it, the rest of Part II is just using the thing.

A warning that applies to this chapter with unusual force: **pricing and tier boundaries are the single most volatile thing about Grok.** The names of the tiers, the dollar amounts, what is bundled into each, the usage limits, and which features unlock where — all of it changes on a scale of weeks to months, and Grok's access structure has been reorganized more than once already. So I am going to describe the *shape* of access, which is reasonably durable, and deliberately **not** print specific prices, because any number I committed to paper here would likely be wrong by the time you read it. For the actual current figures, there is exactly one authoritative source: xAI's own pricing page and X's subscription pages, checked today. Treat everything below as the map, not the territory.

Two Doors Into Grok

The first thing that trips people up is that Grok is sold through two different storefronts, and they are easy to confuse.

Through X. Because Grok was born inside X and is still deeply tied to it, one way to get Grok — including its more capable tiers — is as part of an X subscription. If you already pay for X's premium subscription, you very likely already have a meaningful level of Grok access bundled in, and the upgrade to more Grok is, in effect, an upgrade to your X plan.

Through xAI directly. Grok is also sold on its own, separate from any X subscription, via xAI's standalone offering. You can pay xAI for Grok access and use it on the grok.com website and the dedicated mobile apps without subscribing to X at all.

The practical consequence is that there is no single "Grok price," and two people can be paying for very similar Grok access through entirely different doors. When you go to check current pricing, look at both X's subscription tiers and xAI's standalone tiers, because the best deal for *your* situation depends on whether you also want X. The structure has shifted before and may have converged or split further by the time you read this — another reason to verify rather than trust a remembered arrangement.

A small worked example makes the two-doors problem concrete. Imagine two colleagues who both want "good Grok." One already lives on X — posts daily, follows a few hundred accounts, would happily pay for the platform's premium features regardless of Grok. For her, the X subscription door is almost certainly the right one: the Grok access rides along with features she already wants, and a second standalone Grok bill would be paying twice for overlapping value. The other colleague never opens X, finds the feed actively unpleasant, and wants Grok purely as a working assistant on grok.com. Paying for an X subscription to reach Grok would mean buying a social platform he will never touch just to get the assistant; the standalone xAI door is the obvious fit. Same desired capability, opposite correct answers, and the only variable that flipped was *do you also want X*. Ask yourself that question first and the rest of the tier decision gets noticeably easier.

The Shape of the Tiers

Without pinning numbers that will age badly, here is the ladder as of this writing, from least to most.

Free. There is a free tier, and it is genuinely useful — enough to try Grok, get a feel for the interaction, and run real queries, including a taste of its real-time strength. But it comes with tight usage limits and without the full power of the heaviest features and highest-capability model behavior. It is the right place to *start* and the wrong place to *settle* if Grok turns out to be something you lean on. Think of it as the test drive.

X Premium+. The top tier of X's own subscription has historically been the route to a fuller Grok experience for people who live on X anyway — higher limits and access to more of Grok's capabilities, bundled with the platform's other premium features. If you are an active X user, this is often the most natural door, because you get the social-platform benefits and a strong slice of Grok in one subscription. Exactly how much Grok, and at what limits, is a verify-it number.

SuperGrok. This is xAI's standalone subscription for people who want serious Grok access without necessarily routing it through X. It raises usage limits and opens up the more demanding capabilities — the deeper research modes, heavier creative generation, the most capable model behavior — for someone whose primary interest is Grok itself rather than the X platform. For a lot of professional users who want Grok as a working tool, this is the tier to look at first.

Heavy. At the top sits the highest tier — the one aimed at the most demanding users, with the highest limits, the earliest or fullest access to the most capable model configurations, and the most headroom before you hit a wall. It is priced accordingly, well above the others, and it is meant for people whose use is intensive enough that the ceiling on lower tiers genuinely constrains them. Most individuals will never need it; the heaviest daily users and professionals who run Grok hard all day are who it exists for.

The honest summary, names and numbers aside: **free to try, a mid tier to actually rely on it day to day, and a top tier for people who lean on it hardest** — reachable through either an X subscription or xAI's standalone plans depending on which door suits you. That structure is reasonably stable even as every specific around it moves.

What "More" Actually Buys You as You Climb

The tier names tell you the order; they do not tell you *what changes* as you go up, and that is the part worth understanding before you spend. Broadly, climbing the ladder buys movement along three separate axes, and different people care about different ones.

More usage before you hit a wall. This is the axis everyone notices first. Free tiers cap how much you can do in a given window tightly; each step up loosens that cap. If the thing that drives you to upgrade is "I keep running out," you are buying headroom on this axis, and the question is simply whether your volume justifies the next rung.

Access to the heavier capabilities at all. Some of Grok's most distinctive features — the deepest research runs (Chapter 9), the most demanding creative generation (Chapter 12 and Chapter 13), long sessions in the most capable model configuration with heavy reasoning (Chapter 8) — are either unavailable or sharply limited on the lighter tiers. Here you are not buying *more* of something you already have; you are buying the ability to do a thing you could not do before. For a lot of professional users this axis, not raw volume, is the reason to pay.

The most capable model behavior and the earliest access to it. The top tiers tend to offer the fullest, strongest model configurations and, often, the first crack at new capabilities as they roll out. If being on the leading edge matters to your work, that is what the highest rung is partly selling.

The reason to separate these axes is that they let you diagnose your own situation honestly. "I keep hitting limits" is an argument for the next rung up on the volume axis. "The feature I want is greyed out" is an argument about the capability axis, and may mean jumping further than one rung. "I just want the best Grok can do, cost be damned" is the top-rung argument. Knowing *which* of these is actually pushing you keeps you from over-buying — paying for the top tier when a middle one would have unlocked the single feature you were after, or vice versa.

A Couple of Concrete Profiles

To make the ladder less abstract, here is how the decision tends to fall for a few recognizable people — with the standing reminder that you must confirm current terms before acting.

The journalist who lives on X. She is in the feed all day anyway, fact-checks claims in real time, and wants Grok's live edge constantly. The X-subscription door fits her life, and a higher X tier that bundles a generous slice of Grok is likely her best value. She rarely needs the very heaviest research runs, so the top standalone tier would be money spent on headroom she would not use.

The consultant who runs deep research for client work. He does not care about X as a platform but leans hard on DeepSearch and the most capable model for analysis he will be paid for and held accountable to. SuperGrok-style standalone access is his natural home, and if his research volume is heavy enough that limits start biting mid-project, that is exactly the signal that escalating toward the top tier is worth pricing — not before.

The curious newcomer. No idea yet whether Grok will stick. The only correct first move is Free: run real queries, especially the real-time ones, and let actual use rather than a sales page decide whether any paid tier earns its place. Upgrading later costs nothing you have lost by waiting.

Usage Is Metered Even Within a Tier

A point that surprises people: paying for a tier does not buy unlimited use. Within every tier there are usage limits — caps on how much you can do in a given window — and the more demanding the feature, the faster it draws those limits down. A quick question costs almost nothing. A deep, multi-source research run (Chapter 9), a batch of high-resolution image generations (Chapter 12), a long session in the most capable model with heavy reasoning (Chapter 8) — these consume far more of your allowance, and you can move through a tier's ceiling faster than casual experience would suggest.

This is not cause for alarm; it is cause for awareness. If you find yourself hitting limits, the question is whether you are using the expensive features heavily enough to justify a higher tier, or whether a lighter touch would keep you comfortably inside the one you have. Either way, the meter runs, and the heavy features run it faster.

It helps to have a rough intuition for the relative weights, even without exact numbers. Picture a single afternoon. A dozen ordinary questions — quick lookups, short rewrites, a definition or two — barely register; you could ask those all day on most tiers and never feel a ceiling. One serious DeepSearch run on a sprawling topic is worth many of those quick questions in resource terms, because the model is planning, searching, reading across many sources, and synthesizing. A handful of high-resolution image generations is heavier still in its own way. And a long, multi-turn session in the most capable model with reasoning turned up high is the heaviest of all, because every turn is doing expensive work and there are many turns. The lesson is not to ration nervously — it is to recognize that your "I'm hitting limits" experience will be driven almost entirely by how often you use the three or four heavy features, not by how *much* you chat overall. People who blow through a tier are rarely the people who talk to Grok the most; they are the people who run the expensive modes the most. If you are bumping the ceiling, look at your heavy-feature habits first.

A Separate Door for Builders: the API

Everything above is about using Grok interactively, as a person, through a subscription. There is an entirely separate path for developers: the **API**, where you pay per use — by the token — to call Grok's models from your own software rather than through a subscription tier. The billing model is different (metered usage, not a monthly plan), the capabilities you reach are different, and it is the right path for scripting, automation, and building Grok into applications. If you are an individual getting started interactively, ignore the API for now; Part VI is where it gets its due, and Chapter 22 covers its costs specifically. I mention it here only so you know the subscription tiers are not the whole story — there is a builder's door with its own meter.

The distinction between the two billing models matters more than it first appears, and confusing them is a common early mistake. A subscription is a flat monthly cost for interactive use with usage caps — you pay the same whether you use Grok lightly or right up to the ceiling, and the cap protects xAI from your heaviest days. The API is the opposite shape: there is no monthly floor and no cap, just a price per token that scales directly with how much you call it. That makes the API beautiful for unpredictable or automated workloads and dangerous for the careless, because a runaway loop or an unexpectedly popular feature can run up a bill with nothing to stop it. The rule of thumb: if a human is sitting there typing, a subscription tier is almost always the right and cheaper choice; if software is doing the calling, the API is the path, and Chapter 22 is where you learn to keep its meter from surprising you.

Which Tier Should You Start On?

A practical recommendation, with the standing caveat that you should confirm current terms before acting on any of it:

If you are **curious and unsure**, start on **Free**. Run real queries, especially the real-time ones that show off what is distinctive, and confirm Grok earns a place in your toolkit before paying anything.

If you **already pay for X** and use it regularly, look hard at **X Premium+** first — you may get most of the Grok you want bundled with a subscription you already hold or would value anyway.

If you **want Grok as a serious working tool** and are not especially attached to the X platform, look at **SuperGrok** — it is built for exactly that person, and it opens the demanding features most professionals actually want.

If you are a **power user** who will run deep research, heavy generation, and long high-capability sessions all day, expect to end up evaluating **Heavy** — but do not start there. Move up to it when the lower tiers' limits actually bite, not in anticipation.

And if you are a **developer** planning to build on Grok rather than chat with it, the subscription ladder may be beside the point — go straight to the API path in Part VI.

What You Need Before the Next Chapter

To follow the rest of Part II, you need very little: an account — through X or xAI, whichever door you chose — a decision about which tier to begin on (Free is a fine answer), and a willingness to verify the current prices yourself before you commit, because I have deliberately not put them here and I would be doing you a disservice if I had. Get that settled — an account, a starting tier, and prices you confirmed rather than took on faith — and the spending decisions are behind you. What is left is the good part: actually using the thing, beginning with where Grok lives and which of its surfaces fits which moment.

Chapter 6: Grok on X, Web, and Mobile

I was standing in a checkout line when a friend texted me a screenshot of a post that was apparently blowing up and asked, "is this real or is someone yanking my chain?" I pulled up X on my phone, found the post, tapped the little Grok button in the reply composer, and asked whether the claim in it held up. By the time I reached the register I had an answer that pointed at the original source, flagged that the screenshot had been cropped to drop a "satire" label, and noted who was amplifying it. I never left the app the post lived in. That is the small, ordinary magic of Grok meeting you where the conversation already is — and it is the right place to start a chapter about *where* you actually use this thing.

Because here is a fact that trips up newcomers: Grok is not one place. It is the same assistant reachable through several different front doors, and the door you walk through shapes what the experience is good for. This chapter is a tour of those doors — the grok.com web app, the mobile apps, and Grok living inside X itself — what each is genuinely best at, how they differ, and why the X-native surface is special in a way the others structurally can't match. One standing caveat before we begin: which features live behind which door, and which tier unlocks them, is exactly the kind of thing that drifts month to month. I will describe the durable shape of each surface and point you to Chapter 5 for who-gets-what. **Confirm the current feature-by-surface breakdown against xAI's own pages** before you build a habit on it.

Three Front Doors, One Assistant

The mental model to hold is simple: there is one Grok — the same family of models from Chapter 4 — and there are several surfaces that put you in front of it. The surfaces differ not in *which Grok* you get but in the framing around the conversation: how much screen you have, what's one tap away, and crucially, what context the surface already knows about. A chat on the website starts from a blank page. A chat invoked on an X post starts knowing about that post. Same engine, very different starting line.

So the question is never "which one is best" in the abstract. It is "which door fits what I'm trying to do right now." Let me take them in turn.

grok.com: The Standalone Web App

The website at grok.com is Grok at full desk-bound capability — the surface to reach for when the task is *work* rather than a quick reaction. It is the most spacious experience: a real conversation pane, room to read long answers, easy file and image uploads, and access to the heavier modes — Think mode for hard reasoning (Chapter 8), DeepSearch and DeeperSearch for extended research (Chapter 9) — laid out where you can actually drive them.

This is where I do anything that involves reading carefully or handing Grok a document. Pasting a long report and asking for a structured summary, working through a coding problem across several turns, running a research task and reading the sourced result, drafting something and revising it back and forth — the web app gives you the screen real estate and the persistent conversation history that those jobs want. It is also, for most people, where the full set of capabilities shows up first and most completely, because a browser tab has room for controls a phone notification bubble does not.

If you are sitting down to *use* Grok for something that matters, the website is the default. Treat the other surfaces as specialized tools and this one as the workshop.

To make that concrete, here is the kind of session the website is built for. Say you have a forty-page market report a colleague sent over and a meeting in an hour. You open grok.com, drag the PDF into the conversation, and ask for a structured summary — the three findings that matter, the data behind each, and anything the report glosses over. You read the answer in the wide pane, spot a claim you want to interrogate, and ask a follow-up without re-uploading anything, because the document and the whole thread are still in context. You ask it to draft three talking points for the meeting, then revise the second one to be blunter. Then you switch on DeepSearch (Chapter 9) to check whether anything material has happened since the report was written, and read the sourced result alongside the summary. Every one of those moves — the upload, the long read, the multi-turn refinement, the heavy mode, the persistent thread — is more comfortable with a full screen and a real keyboard. That whole arc is awkward to impossible on a phone notification bubble, and effortless in a browser tab. That is what "the workshop" means in practice.

The Mobile Apps: Grok in Your Pocket

The iOS and Android apps are the same assistant sized for a phone, and their advantage is not capability — it is *presence*. Grok is with you in the line, on the walk, in the passenger seat, at the moment a question actually occurs to you rather than the moment you next get to a keyboard. That changes which questions you ask. Voice mode (Chapter 14) belongs here more than anywhere: talking to Grok hands-free, out loud, is a mobile-native interaction, and it turns the assistant into something you converse with rather than type at. The camera matters too — point your phone at a thing, a sign, a page, a chart, and ask about it, and you are using Grok's vision (Chapter 15) in the most natural possible way.

A few mobile-native moments are worth naming because they are the ones that change your habits. You are in the grocery store holding a jar with an unfamiliar additive on the label; you point the camera at it and ask what it is and whether it matters. You are on a walk and a question occurs to you that you would have forgotten by the time you got home; you say it out loud to voice mode and get an answer before the thought evaporates. You are in a foreign restaurant staring at a menu in a language you do not read; you photograph it and ask what each dish is. None of these are tasks you would ever sit down at a desk to do — the whole point is that the assistant was *there*, in your hand, at the moment the need appeared. That immediacy is a different kind of value from the website's depth, and it is why the apps are not a lesser version of grok.com but a genuinely different tool.

What the apps trade away is room. Long documents and deep multi-turn research are simply more comfortable on a larger screen, and some of the heaviest controls are easier to reach in a browser. The honest division of labor: **the apps are for capture and conversation in the moment; the website is for sustained work.** Most people who use Grok seriously end up using both, and the conversation history syncs across them, so a question you start on your phone you can finish at your desk. That sync is more useful than it sounds: the natural rhythm becomes capturing a thing on mobile — a photo, a quick question, a voice note of an idea — and then picking the thread back up at the keyboard when you have the screen to do real work with it. The phone is where the question is born; the desk is where it grows up.

Grok Inside X: The Surface Only Grok Has

Now the one that makes Grok *Grok*. Because xAI and X are joined at the hip, Grok lives natively inside X — and this is not a bolted-on chatbot link, it is a genuinely different way to interact, built around the thing X is: a live stream of human conversation.

There are a few distinct moves here, and they're worth naming separately.

Asking @grok in public. You can mention @grok in a reply, the way you'd tag a friend, and ask it something — "@grok is this true?", "@grok what's the context here?", "@grok summarize this thread" — and it answers in the thread, in the open, where everyone in the conversation can see. This is fact-checking and context-fetching woven directly into the social fabric. It is also, by design, public: you are asking in front of an audience, which is wonderful for settling a debate in the replies and worth remembering when the question is something you'd rather keep private.

Analyzing a specific post or thread. Invoke Grok on a particular post and it starts with that post already in hand. You don't paste anything or describe what you're looking at — the surface hands Grok the context. "Is the claim in this real?" "What's the original source?" "Walk me through what this thread is actually arguing." The checkout-line moment that opened this chapter was exactly this: the post was the input, and I supplied only the question.

Reaching live reaction at the source. This is where the X surface compounds with Grok's defining edge. When you ask Grok inside X about what's happening, you are asking the assistant that has the firehose of reaction *and* you are asking it from inside the stream that reaction is flowing through. For "what are people saying about this right now," there is no closer place to stand. The real-time advantage that Chapter 1 made the center of the whole book is at its most immediate here.

The public-versus-private fork. Worth pausing on, because it catches people. The headline move — mentioning @grok in a reply — is *public by construction.* Your question and Grok's answer sit in the thread for anyone to read, which is exactly what you want when you are settling an argument in front of an audience or adding context a whole conversation benefits from, and exactly what you do not want when the question reveals something about you, your work, or someone else. Asking "@grok, is the medical claim in this post accurate?" in public is fine. Asking "@grok, given my company's situation, is this rumor about us true?" in public is a mistake you cannot take back. The instinct that protects you is simple: before you mention @grok in the open, ask whether you would be comfortable with the question itself, not just the answer, being permanently visible. When the answer is no, that is a website conversation, not an X reply. The platform usually offers more private ways to interact with Grok too, but the public mention is the one with teeth, so it is the one to be deliberate about.

The trade-off is the mirror of the website's: the X surface is unmatched for live, social, in-context questions and not the place for long private documents or sustained heavy work. You would not paste a confidential report into a public reply. Use it for what it's uniquely good at — the live, the contextual, the in-the-stream — and step over to the website when the job changes shape.

Tier Shapes What You See

A wrinkle that catches people, and the reason this chapter keeps pointing at Chapter 5: what's available on a given surface often depends on your tier. The heaviest modes, the highest usage ceilings, and certain features can be gated, and the gating differs by door — something reachable on the website under one subscription may look different inside X, and the free experience necessarily offers less than the paid ones. The tier structure (Free, X Premium+, SuperGrok, Heavy) is real, but I'm deliberately not pinning consumer prices here; Chapter 5 maps exactly what each tier unlocks. The point for now is just to expect that *where* you reach Grok and *what plan* you're on jointly determine what you can do — and that both move, so verify against current docs rather than against this page.

Moving Between Doors Mid-Task

The most fluent way to use Grok is to stop thinking of the surfaces as separate apps and start thinking of them as one assistant you hand off between. A realistic chain: you are scrolling X and a post claims something startling about a company you follow. You tap Grok right there in the stream and ask whether the claim holds — the in-context, in-the-moment move the X surface is built for. The answer is interesting enough that you want to actually understand the situation, not just spot-check the post. So you carry the question over to grok.com, where you have room to run a DeepSearch, read the sourced result, and ask the follow-ups that a real investigation needs. Later, away from your desk, a related thought strikes you on a walk, and you raise it with voice mode in the app — and because history syncs, you are continuing the same line of inquiry, not starting over.

That hand-off is the skill. Each door did the part it was best at: X for catching the live thing in context, the website for the heavy sustained work, the app for capture on the move. You never fought a surface to do something it was wrong for, and the thread followed you across all three. Once that becomes reflex, the question "which door?" stops being a decision you labor over and becomes something you feel — you reach for the one that fits the moment, and switch the instant the moment changes.

Picking the Door

Strip it to a habit you can actually carry. Reach for the **website** when you're sitting down to work — long inputs, deep research, careful multi-turn reasoning, anything you'd want a full screen for. Reach for the **app** when the question happens in the world — on the move, out loud, through the camera, in the moment. Reach for **Grok inside X** when the question is about X itself — a post, a thread, a live reaction, a "is this real" that the conversation already contains.

That last door is the one no competitor can copy, because it requires owning the stream. A frozen assistant can be handed a screenshot of a post; only Grok-in-X is already standing where the post lives, with the live reaction flowing past. The checkout line worked because I never had to leave the conversation to ask about it — and once you've felt that, the other surfaces make sense as what they are: the same capable assistant, fitted to a different moment in your day.

Chapter 7: Your First Conversations

The first thing I ever asked Grok, sitting at my desk with the website open and a slightly skeptical eyebrow raised, was a softball: "what are people saying about the game last night?" I picked it because I genuinely didn't know — I'd missed the game — and because it was the kind of question that would have stopped any frozen assistant cold. Grok answered in a breath: the result, the moment everyone was talking about, the split between the fans who were elated and the ones who were furious about a call, a couple of the sharpest posts quoted. It was right. I checked. And the eyebrow came down, because in about fifteen seconds the abstract promise of "real-time AI" had become a concrete thing that had just worked for me.

That is the whole purpose of this chapter: to get you from reading about Grok to having a real first session, so that the capability stops being a claim and starts being something you've done. Reading about it convinces no one. One good first conversation convinces everyone. So let's have one.

Just Start Talking

The good news is that there's almost nothing to learn before you begin. Grok is a conversation, not a command line — you type what you want in plain language, the way you'd ask a sharp, fast colleague, and it answers. There is no special syntax, no keywords to memorize, no wrong button. Open the website or the app, find the text box, and type a real question. The single most common mistake I see in first-timers is overthinking the *first* message, treating it like they have to phrase a search query just so. You don't. Ask the way you'd ask a person.

So pick something you actually want to know and send it. If you need a nudge, here are first questions that tend to land well — concrete, useful, and a fair test of the tool:

Something from your real life. "I have chicken thighs, rice, and whatever's usually in a pantry — give me three dinner options and pick the best one." A request with real constraints shows you how Grok handles specifics.

Something you'd otherwise google a dozen tabs for. "Explain the difference between a Roth and a traditional retirement account like I'm smart but new to this." Watch how it pitches the explanation to the level you asked for.

Something current — this is the one to try first. "What's the reaction so far to [whatever's in the news today]?" This is the question that shows you what's different, and we'll come back to it.

Send one. Read the answer. You've now had a conversation with Grok, which is more than most write-ups will get you to do.

What you'll notice on that first answer, whatever you asked, is that Grok writes back in full prose, not bullet-pointed search results — it's talking to you, not returning hits. If you asked the dinner question, it didn't just list recipes; it weighed your constraints, made a recommendation, and probably told you *why* it picked the one it did. That "and here's my reasoning" texture is worth noticing on day one, because it's the thing you'll lean on later: you can always ask it to defend or reconsider a choice, and it will.

Vague In, Vague Out

Now the one idea that will improve your results more than any other, and it's not a trick — it's a habit. **The quality of what you get back tracks the quality of what you put in.** A vague request gets a vague, hedge-everything answer because Grok is guessing at what you meant. A request with specifics gets a specific answer because there's nothing left to guess.

Watch the difference. A weak request: *"give me some tips on email."* Tips for whom, about what, toward what end? Grok will produce something generic because you handed it nothing to aim at. A strong request: *"I'm a freelance designer and my follow-up emails to prospects go unanswered. Rewrite this one to be warmer and to make the next step obvious — here's the email."* Now there's a role, a problem, a goal, and the raw material. The answer comes back sharp because the question was.

The pattern under that is worth internalizing, because it generalizes to everything you'll ever ask an assistant:

Say who you are or who it's for when it changes the answer — "explain this to a beginner," "I'm a nurse," "this is going to executives."

Say what you actually want — a summary, three options, a rewrite, an explanation, a yes-or-no with reasons. Naming the shape of the output you want is half the battle.

Give it the material and the constraints — paste the text, name the budget, state the deadline, list the things to avoid. Grok can't account for a constraint you kept in your head.

You don't have to do all three every time, and short questions are completely fine for simple things. But when an answer comes back mushy, nine times out of ten the fix isn't a cleverer prompt — it's a more specific one. Add what you left out and ask again.

Watch a Vague Request Get Better in Three Passes

Because this is the single habit that pays off most, it's worth watching it work rather than just reading the principle. Take a real one. You want help with a difficult message to your landlord about a repair that's been ignored.

The first instinct is something like: *"Help me write an email to my landlord."* Grok will produce a perfectly competent generic email — and it will be generic, because you told it nothing. It doesn't know what the repair is, how long it's been, whether this is the third request or the first, or whether you want to sound firm or stay friendly because you'd like the lease renewed.

So you add the role and the goal: *"I'm a tenant. The heating's been broken for two weeks despite two prior requests. I want to be firm and signal I know my rights, but not hostile — I'd still like to renew the lease."* Now the answer changes shape entirely. It's firmer, it references the prior requests, it leaves the door open. Already usable.

Then you give it the constraints and material that only you have: *"Keep it under 150 words. Mention that I've documented the dates of every request. Don't threaten legal action outright, just imply I'm aware of the option."* The third answer is the one you actually send — tight, specific, calibrated to exactly the tone you wanted. Three short passes got you somewhere a single perfect prompt rarely does, and notice that none of the three required clever wording. Each just *added what was missing*. That's the whole move, and it generalizes to nearly everything you'll ever ask.

It's a Conversation, So Keep Talking

Here's the thing first-timers most often miss: you are not allowed only one shot. Grok remembers the thread, so the most powerful move in your toolkit is the follow-up. The first answer is a draft you steer, not a verdict you accept.

So push on it, the way you would with a person who just told you something. "Shorter." "More formal." "You missed the part about the deadline — redo it with that." "Why did you recommend the first option over the second?" "Give me three more like the second one." Each turn refines, and because Grok holds the context of everything said so far (Chapter 4 covered just how much it can hold), you don't have to re-explain — you just nudge. I get to a good result far more often by sending a rough first request and refining it twice than by trying to write one perfect prompt up front. Let the conversation do the work.

A few follow-up moves are worth knowing by name because they're the ones you'll reach for constantly:

The reshape. "Same content, but as a bulleted list." "Turn that into a short paragraph instead." You're not asking for new thinking, just a different container for the thinking it already did.

The challenge. "What's the weakest part of that argument?" "Push back on what I just said — where am I wrong?" Grok will too readily agree with you if you let it; explicitly asking it to disagree is how you get the dissent you actually need.

The why. "Why did you choose that over the alternative?" Surfacing its reasoning lets you check whether the answer rests on something solid or something you'd reject if you saw it.

The fork. "Give me three more options in the style of the second one." When one direction is promising, this multiplies it without starting over.

The correction. "No — I said the deadline is Friday, not next week. Redo it." Grok takes correction cleanly and folds it into everything downstream. You don't have to be polite about it or re-explain the whole task; just point at what's wrong.

None of these are advanced. They're the ordinary moves of talking something through with a capable colleague, and the reason to name them is that first-timers often don't realize they're allowed. You are. The first answer is raw material, and steering it is where most of the value lives.

Try the Thing Only Grok Does

Everything so far would be true of any capable assistant. This part is where Grok shows you why it exists, and you should try it in your very first session, because it's the fastest way to feel the difference in your gut.

Ask it something about *right now*. Not last year — this morning. Try a handful and watch what comes back:

"What's the reaction to [today's news] so far?" This was my first question for a reason. A frozen assistant apologizes; Grok goes and reads the live reaction and tells you the mood, the split, the sharpest takes.

"What are people saying about [a product, a show, a team] today?" Live sentiment, drawn from what's actually being posted, not reconstructed from memory.

"Did [a thing you've been waiting on] actually ship/happen yet?" A frozen model can't know; Grok can go check.

"What's trending in [a topic you follow] right now?" A read of the live conversation in your corner of the world.

When the answer leans on live sources, get in the habit early of glancing at where it came from — Grok aims to show you, and the live web carries rumor and noise alongside truth (Chapter 1 flagged this; Chapters 23 through 26 treat it honestly). But the *feeling* of asking about the present and getting a real answer is the whole point of trying these immediately. It's the moment the skepticism falls away.

Here's a small test that makes the difference unmistakable, and it costs you nothing. Pick something you genuinely don't know the current state of — a film that just came out, a phone that was rumored to be launching, a sports result from last night, the reaction to some announcement in your field. Ask the same question two ways. First, ask it cold, from general knowledge — and notice that for anything truly recent, even Grok would be guessing. Then ask it to actually check what's being said right now. The second answer comes back grounded in live material, often with the split of opinion and the sharpest takes, and frequently with sources you can click. Run that comparison once on a topic you'll be able to verify yourself, and the abstract claim of this whole book — that real-time is a different capability, not a marketing line — becomes something you've personally watched happen. That's worth more than any number of pages telling you so.

A Few Honest First-Session Notes

So your expectations are calibrated and not just inflated: Grok is fast, fluent, and broadly capable, and it can also be confidently wrong. A smooth, assured answer is not automatically a correct one — fluency is not accuracy, and that's true of every model in the field. For anything that matters — a number you'll act on, a fact you'll repeat, a claim you'll stake something on — verify it, especially when it rests on live sources you can go check yourself. This isn't a knock on Grok; it's the basic posture for using any assistant well, and it's cheap insurance.

You'll also notice Grok answers easy things instantly and can take longer on hard ones — that's the reasoning dial from Chapter 4 doing its job, and Chapter 8 is entirely about driving it. For now, just don't be surprised when a genuinely hard question makes you wait a beat; that pause is Grok thinking, and it's usually time well spent.

A couple of other things will feel slightly off at first and are worth knowing in advance so they don't throw you. Grok can be agreeable to a fault — if you assert something confidently, it will often go along with it rather than correct you, so the burden of skepticism stays with you, and the "push back on me" move above exists precisely to counter that tendency. It can also occasionally over-explain, padding a simple answer with caveats and context you didn't want; the fix is just to tell it ("shorter," "skip the preamble," "just the answer"), and it'll recalibrate for the rest of the thread. And on the real-time questions especially, it can present something it read on the live web with more confidence than the source deserves — which is the whole reason the next caution exists. None of these are dealbreakers. They're the normal rough edges of a fast, fluent assistant, and a session or two is enough to learn to steer around them.

Go Have the Conversation

If you take one thing from this chapter, let it be that there's no preparation required and no way to do it wrong — there's just the conversation, which gets better the more specific you are and the more you follow up. Pick a real question, ask it plainly, and then ask about this morning's news to feel the part no other assistant does. Most of the rest of this book is refinement on top of that loop: ask, read, refine, and verify what matters. You already know how to do it. Open the box and say something.

Chapter 8: Think Mode and Reasoning

I once handed Grok a logic puzzle that had been bugging a group of us at dinner — one of those who-sits-where, the-doctor-isn't-the-tallest constraint puzzles where every clue depends on three others. Asked plainly, the first answer came back fast and confidently wrong; it had picked an arrangement that violated a clue four lines up. So I asked it again, the same puzzle, but this time told it to work through it step by step. It slowed down. It laid out the people, the seats, the constraints, eliminated possibilities one at a time, caught the contradiction the fast answer had blown past, and arrived — correctly, this time — at the only arrangement that fit. Same model, same question. The only thing that changed was that I let it *think* before it answered.

That difference is the subject of this chapter. Grok can reason in two registers — a fast, direct one for ordinary questions and a deliberate, step-by-step one for hard ones — and knowing when to invoke the deliberate mode, what it costs you, and when it's just making you wait is one of the most practically useful skills in the whole book. Chapter 4 introduced this as the reasoning dial built into the Grok 4.x generation. Here's how to actually drive it.

What Think Mode Actually Is

When you ask Grok an easy question, it answers the way you'd answer "what's the capital of France" — immediately, off the top of its head, no deliberation required. **Think mode** is the other gear. In it, Grok works the problem out explicitly before committing to a reply: it breaks the question into parts, plans an approach, reasons through the steps, checks intermediate results, and catches its own mistakes along the way, rather than blurting the first plausible answer.

The useful mental image is the difference between a snap reply and showing your work. A fast answer is a confident first instinct, which is exactly right for easy questions and dangerous for hard ones — because on a hard problem, the first plausible-sounding answer is often a trap, and only working through the steps reveals the contradiction. Think mode trades speed for that working-through. It is the model planning and checking instead of guessing, and on the right problem it is the difference between an answer that sounds right and one that *is* right.

This isn't a bolted-on gimmick. The capacity for extended, deliberate reasoning is a defining feature of the current model generation (Chapter 4), and it reflects where frontier AI has landed: the most capable systems can now spend more computation on harder problems, and the gain is real.

Watching the Two Gears Diverge

The clearest way to feel the difference is to watch the two registers hit the same problem and split apart. The dinner puzzle is one shape of this; here's another that comes up in ordinary life.

Suppose you ask: "I'm 41, I want to retire at 62, I can save $1,200 a month, and I expect roughly 6% average annual returns — roughly where do I land, and is that enough to replace $5,000 a month in income?" Ask that of the fast gear and you may get a confident number that quietly skips a step — maybe it computes the savings total but forgets to ask whether that total, drawn down over a multi-decade retirement, actually throws off $5,000 a month. The answer *sounds* complete. It isn't.

Now ask the same question and tell it to work through it step by step. It slows down and lays the problem out: twenty-one years of contributions, compounding at the assumed rate, to reach an accumulated figure; then a separate question of what sustainable monthly income that figure supports over a retirement of some length; then a comparison against the $5,000 target, with the gap named explicitly. Crucially, the careful version is more likely to *flag its own assumptions* — that 6% is optimistic after inflation, that the drawdown math depends on how long you live, that taxes haven't been touched. Same model, same question. The fast gear gave you a number; the slow gear gave you an analysis, and on a question where a skipped step changes the conclusion, the analysis is what you actually wanted.

The pattern under both examples is the same: hard problems have *dependencies* — later steps rest on earlier ones — and the fast gear is prone to satisfying the surface of the question while violating a dependency it didn't pause to check. Deliberate reasoning exists to honor those dependencies.

When It Earns Its Keep

Think mode is not free — it costs you time, and sometimes money (more on both shortly) — so the skill is knowing when the trade is worth it. The honest rule is the one that governs all of modern AI: **think harder for hard things, answer fast for easy ones.** Concretely, reach for deeper reasoning when the problem has one of these shapes:

Multi-step logic where one wrong step poisons the whole answer. The dinner puzzle. Constraint problems, deductions with several interlocking conditions, "if A then B but not C" chains. Anything where a fast guess is likely to violate something stated earlier.

Math beyond the trivial. Not "what's 17 times 4" — Grok handles that directly — but multi-step word problems, anything requiring careful setup, a calculation where the path matters and a skipped step changes the result. Deliberate reasoning dramatically lowers the rate of confident arithmetic mistakes.

Genuine analysis with competing considerations. "Here are three vendors with different pricing, terms, and risk profiles — which fits a cash-tight startup, and why?" When the answer requires weighing factors against each other rather than retrieving a fact, the step-by-step gear produces a sturdier, better-justified result.

Tricky code. Debugging logic that doesn't behave, reasoning about edge cases, designing an algorithm where correctness isn't obvious by inspection. Chapter 18 goes deep on coding; the relevant point here is that hard code is a reasoning problem, and reasoning problems are what Think mode is for.

Planning with interacting constraints. "I have four things to ship this quarter, two of them depend on the same engineer, one has a hard external deadline, and I'm losing a week to a conference — sequence them." This is the work-life cousin of the logic puzzle: the constraints interact, and a fast answer tends to produce a plausible-looking order that quietly breaks one of them. The deliberate gear actually checks the order against each constraint.

Anything where you'd want a smart person to slow down. That's the gut check. If you'd be nervous taking the answer from someone who replied in half a second, that's a Think-mode question.

A quick way to triage in the moment: ask yourself whether the answer is *retrieved* or *derived*. Retrieved answers — facts, definitions, recollections, summaries of something you handed it — live in the fast gear and gain nothing from deliberation. Derived answers — anything the model has to build by reasoning across several steps, where the steps depend on each other — are where Think mode earns its keep. "What's the boiling point of water at altitude" is retrieved-ish and fine fast. "Given these four constraints, what's the optimal choice" is derived, and that's a slow-gear question.

When It Just Makes You Wait

The flip side matters as much, because over-reaching for Think mode is a real and common mistake. On easy questions, deep reasoning buys you nothing but latency. A simple factual lookup, a quick definition, a short rewrite, "what's the reaction to today's news," a casual back-and-forth — these don't have hidden steps to work through, so making the model deliberate over them just makes you sit and wait for an answer the fast gear would have nailed instantly. Worse, on the kinds of questions where being *current* is the whole point — the real-time, live-knowledge questions that are Grok's signature — what you usually want is a fast read of fresh sources, not a long internal deliberation. Think mode is for hard reasoning, not for freshness; don't confuse the two.

The cost isn't only your time. Deeper reasoning does more work under the hood, and on metered usage — API calls especially, where you pay for the tokens the model generates, and reasoning generates more of them — that work shows up on the bill. Chapter 22 covers the cost mechanics in detail. The durable point: reasoning is more expensive in both seconds and tokens, so spend it where the difficulty actually lives and not by reflex.

There's a subtler over-reach worth naming too: invoking deep reasoning on a problem that *feels* hard but isn't actually a reasoning problem. "Write me a heartfelt toast for my sister's wedding" feels weighty, but it's a writing task, not a deduction — the slow gear won't make it more moving, just slower to arrive. "Summarize this long article" feels like a lot of work, but it's comprehension, not multi-step logic; the fast gear handles it fine. The tell isn't how *important* or how *long* the task is — it's whether getting it right requires chaining steps where one wrong step breaks the rest. Importance and length fool people into reaching for the heavy gear; only genuine step-dependency justifies it.

How to Invoke It and What to Expect

The exact control — a button, a toggle, a mode selector, or in some cases Grok deciding on its own that a question warrants deliberation — varies by surface and shifts over time, so **check the current interface rather than memorizing a specific button.** What's durable is the intent: you can ask for, or switch into, deliberate reasoning when a problem needs it. In a typed conversation you can also simply ask — "think this through step by step before you answer" is a request the model understands and will honor, and it's a reliable way to get the careful gear even when you're not sure where the toggle is.

Expect three things when you do. First, it's slower — sometimes noticeably, because working through a hard problem takes real steps, and that pause is the feature, not a stall. Second, you often get to see some of the reasoning, which is genuinely useful: you can follow how it reached the answer and catch a wrong turn in the logic, which is far better than auditing a bare conclusion. Third, the answer is usually better on exactly the problems you invoked it for — and no better, just slower, on the problems you shouldn't have. That last asymmetry is the whole lesson in miniature.

Reading the Reasoning, Not Just the Answer

That second point — getting to see the work — deserves more than a clause, because it changes how you use the mode. When Grok shows its reasoning, you are not meant to skim past it to the bottom line. The visible steps are where you catch the errors that a bare conclusion would hide.

Concretely, scan the reasoning for three things. **Did it use the right inputs?** On the retirement question, did it actually plug in $1,200 a month and 6%, or did it drift to round numbers somewhere? A wrong input early poisons everything after it, and you can only see that if you read the early steps. **Did it honor every constraint?** On the quarter-planning question, walk the proposed sequence against your four constraints yourself — the reasoning makes that checkable in a way a bare ordering never could. **Is there a leap?** Watch for the place where it jumps from "here's the situation" to "therefore" without the connective step. That leap is where confident-but-wrong lives.

This is also the honest limit worth stating plainly: the visible reasoning is a genuine aid, but it is not a guarantee. A model can produce reasoning that looks sound and still reach a wrong answer, and it can occasionally arrive at the right answer by a route that doesn't actually hold up. Seeing the work makes errors *easier to catch*, not impossible to make. So read the steps as a colleague's scratchpad you're sanity-checking, not as a proof you can rubber-stamp. The reasoning earns trust; it doesn't replace yours.

What Reasoning Can't Fix

It's worth being clear about the boundary, because over-believing in Think mode is its own trap. Deliberate reasoning makes the model better at *working through* a problem. It does nothing to fix bad or missing inputs. If you hand it a constraint that's wrong, it will reason flawlessly to a wrong conclusion. If the question secretly depends on a fact the model doesn't have, slowing it down won't conjure the fact — it'll just produce a more elaborate guess. And reasoning is not the same as currency: on a question whose answer depends on something that happened this morning, no amount of internal deliberation substitutes for going and reading the live sources, which is what DeepSearch (Chapter 9) is for. Think mode sharpens the model's *thinking*; it can't supply *knowing*. Match it to problems where the difficulty is genuinely in the reasoning, not in the facts, and it delivers. Aim it at a freshness problem or a missing-information problem and you've just bought a slower wrong answer.

A Habit, Not a Setting

Don't treat Think mode as a switch you flip on and forget, or one you avoid out of impatience. Treat it as a judgment you make per question, and the judgment is easy once you've internalized the shape: hard reasoning, deep analysis, multi-step logic and math, tricky code — slow down and let it think. Quick facts, fresh news, simple rewrites, casual chat — let it answer fast. When you're genuinely unsure, the cheap experiment is to ask once fast, and if the answer feels shaky or the problem is clearly intricate, ask again with reasoning and compare. The dinner puzzle taught me that in one evening: the fast answer wasn't a worse model failing, it was the right model in the wrong gear. Match the gear to the road, and Grok's reasoning becomes one of the most dependable things it does.

Part III: Real-Time and Search

Audience: everyone. This is the heart of the book — the capability that makes Grok a different tool rather than a differently branded one.

"The present is the one thing a frozen model can never reach. This is the part of Grok built to live there."

— A working principle

Chapter 9: DeepSearch and DeeperSearch

I once needed to understand a regulatory change in an industry I knew almost nothing about, and I needed it before a call that started in twenty minutes. The rule had been finalized days earlier. There were a handful of news write-ups, a long official document nobody sane would read end to end in twenty minutes, a trade group's reaction, and a lot of confused arguing online about what it actually meant. I typed the question into Grok, switched on DeepSearch, and went to refill my coffee. When I came back there was a synthesized brief on the screen: what the rule changed, who it affected, the two competing interpretations of the ambiguous clause, and which sources said what — each claim with a link I could click. It was not perfect. But it was a genuinely good starting point built from current material in the time it took to make coffee, and I walked into the call able to ask intelligent questions instead of apologizing.

That is what this chapter is about. DeepSearch and its heavier sibling DeeperSearch are the features that turn Grok's real-time edge from a party trick — "tell me what's happening right now" — into an actual research workflow. They are the flagship of Part III, and the place where living in the present stops being a slogan and becomes a tool you do real work with.

What These Modes Actually Are

A plain Grok answer, even a good one, mostly draws on what the model already knows plus maybe a quick look at a source or two. That is fine for most questions. But when a question is current, contested, or sprawling — when the honest answer requires going out and *reading* a pile of live material and making sense of it — a quick answer is not enough. That is the gap these modes fill.

DeepSearch is an extended, multi-source research mode. When you invoke it, Grok does not fire back an answer off the top of its head. It plans an investigation, searches the live web *and* X, pulls from many sources rather than one, reads what it finds, and hands you back a synthesized answer with the sources attached. The output is not a recollection and not a single-source summary — it is a fresh read of current material, assembled into something coherent, with the receipts so you can check it. Think of it as the difference between asking a knowledgeable friend a question and asking a researcher to spend half an hour and come back with a briefing.

DeeperSearch is the heavier variant of the same thing. Same basic idea — plan, search broadly, read, synthesize, cite — but turned up: more sources, more thorough digging, more cross-checking, and correspondingly more time to run. Where DeepSearch might consult a solid spread of sources, DeeperSearch goes wider and pushes harder on the corners, following threads DeepSearch would reasonably stop short of. You pay for that in latency. A DeeperSearch run is something you start and walk away from, not something you wait on with the cursor blinking.

The exact names, the exact way each mode is labeled in the interface, and which tier gives you how much of each are the kind of details that drift — **confirm the current branding and any usage limits against xAI's own documentation** (Chapter 5 covers tiers). The durable thing is the shape: a lighter real-time research mode and a heavier one, both producing sourced synthesis rather than a quick guess.

How They Differ From a Plain Answer

It is worth being precise about what changes when you switch these modes on, because the difference is not just "more effort."

A plain answer optimizes for speed and draws heavily on what the model already carries. It is the right tool for "explain this concept," "rewrite this paragraph," "what's a good approach to this code." None of those require going out and reading the live web.

A DeepSearch answer optimizes for *currency and breadth.* It assumes the thing you are asking about lives partly or wholly outside the model's training — in this week's news, in today's posts, in a document that did not exist when the model was trained — and it goes and gets it. The model's intelligence is still doing the work of making sense of what it finds; the search is supplying the raw material the model could not otherwise have.

The practical tell for which to use is simple. **If the answer depends on anything recent, contested, or spread across many places, reach for DeepSearch.** If the answer is something a smart, well-read person could give you from what they already know, a plain answer is faster and just as good. Using DeepSearch on "what's the capital of France" wastes your time; using a plain answer on "what's the current state of this fast-moving story" gets you a confident reconstruction of a past that may no longer hold.

What Actually Happens While You Wait

It helps to picture the run, because understanding the mechanics tells you how to write a good DeepSearch query and how to read the result skeptically. When you invoke it, Grok does something closer to what a competent human researcher would do than to what a search box does.

First it *decomposes* your question — turns "what's going on with this regulatory change" into the sub-questions a real answer needs: what exactly changed, when, who's affected, what the reactions are, where the disagreement lies. Then it *searches*, not once but repeatedly, across the live web and X, gathering candidate sources for each sub-question. It *reads* what it pulls, discarding the irrelevant and the duplicative. It *cross-references*, noticing where two sources agree and where they clash. And finally it *synthesizes* — writes up a coherent answer that reflects the spread of what it found, with sources attached so the assertions are traceable.

The reason this matters for *you* is that the quality of the decomposition depends partly on how you framed the question. A vague prompt — "tell me about the new rule" — leaves Grok to guess at the sub-questions, and it may guess wrong, chasing angles you don't care about. A framed prompt — "what did the new rule change, who does it hit hardest, and what are the main competing interpretations of the ambiguous part" — hands it the decomposition and gets you a tighter, more relevant brief. DeepSearch rewards specificity the same way a plain conversation does (Chapter 7), and for the same reason: you're telling it where to aim.

When to Reach for Each

DeepSearch is the everyday workhorse of the two. Reach for it when you want a current, sourced read on something and you are not trying to write a dissertation:

A breaking or recent story you need to actually understand, not just glimpse — what happened, who is affected, what the reaction is.

A topic that touches the last few days or weeks, where the model's training is stale and you need live material woven in.

A question with multiple credible sides, where you want the spread of views and the sources behind them rather than one flattened answer.

A first pass at unfamiliar territory — the twenty-minutes-before-a-call situation — where a sourced briefing beats a blank page.

DeeperSearch earns its extra time for the smaller set of jobs where thoroughness is the whole point:

High-stakes questions where missing a source could cost you — a decision, a public claim, a recommendation you will be held to.

Genuinely sprawling topics with many threads to chase, where DeepSearch's good-enough breadth leaves real gaps.

Cases where you suspect the easy answer is wrong and you want the model to dig past the surface consensus and surface the dissent or the caveats.

The honest guidance is that most of the time DeepSearch is enough, and DeeperSearch is what you escalate to when the stakes or the sprawl justify the wait. Do not default to the heavy mode out of a vague sense that more is better; you will mostly just wait longer for an answer that was not meaningfully more useful.

Three Jobs, Three Different Runs

To make the choice concrete, here is how the two modes earn their place across genuinely different tasks.

The pre-call briefing. The twenty-minutes-before situation from the chapter opening. You need a competent orientation on something unfamiliar, fast, and you'll be using it to ask intelligent questions rather than to make an irreversible decision in the next ten minutes. This is the heart of DeepSearch's range: current, sourced, good-enough breadth, delivered fast. DeeperSearch would be overkill — by the time it finished, your call would have started.

The thing you're going to say in public. You're about to publish a claim, brief an executive, or make a recommendation your name is attached to. Here the cost of missing a source isn't an awkward moment on a call — it's being wrong in front of people who'll remember. This is exactly where DeeperSearch's extra digging and cross-checking pays for its longer wait, because the wider net lowers the odds that some load-bearing source or contradicting view simply never surfaced.

The suspicion that the obvious answer is wrong. Sometimes you start a search already doubting the surface consensus — the popular take feels too clean, or you suspect the loud version of the story is missing something. DeeperSearch is built for this: you specifically want it to push past the easy answer, chase the dissent, and surface the caveats that a quicker run would reasonably skip. Telling it as much in the prompt — "I think the common take here is incomplete; find what it's leaving out" — aims the extra effort where you want it.

Across all three, notice the deciding variable isn't how *interesting* the topic is — it's the cost of an incomplete answer. Low cost, DeepSearch. High cost or active suspicion, DeeperSearch. Curiosity alone doesn't justify the wait.

What a Good Sourced Result Looks Like

The output of these modes has a recognizable anatomy, and learning to read it is part of using them well.

A strong result is **synthesized, not stitched.** It does not just dump five summaries next to each other; it tells you what the sources collectively establish, where they agree, and where they diverge. When two sources conflict, a good result says so rather than silently picking one.

It is **sourced at the claim level**, or close to it. The valuable thing is not a bibliography at the bottom — it is being able to trace a specific assertion back to where it came from, so that when one claim looks shaky you can go check that one without re-doing the whole search.

It **distinguishes what is established from what is reported or rumored.** On a current story, some things are confirmed and some are "people are saying." A good result keeps those separate, and the better it is at that, the more you can trust it. (This is the same noise problem that haunts all live information, treated head-on in the next chapter and again in Chapter 23.)

And it is **honest about its own limits** — flagging where the picture is still developing, where sources are thin, where the answer is provisional. A result that sounds equally confident about everything is a result to be more careful with, not less.

The Ways It Goes Wrong

A research mode is not a truth machine, and the failures have characteristic shapes worth recognizing before they catch you.

It can inherit the bias of what's loud. DeepSearch reads what's available, and on a contested topic the most-posted view is not necessarily the most correct one. If a wrong or one-sided narrative dominates the live conversation, a synthesis built from that conversation can reflect the dominance rather than the truth. The result may *feel* balanced because it cites many sources, while those sources all echo the same origin. The defense is to notice when the breadth is illusory — many links, one underlying claim.

It can dress a rumor in citation. Attaching a source to a statement makes it look verified, but the source might itself be unverified — a post asserting something, cited faithfully, is still just a post. A good result distinguishes "confirmed" from "people are saying," but you cannot assume it always will, and the citation furniture can lend unearned authority to something flimsy. When a striking claim traces back only to posts, treat it as a rumor with a footnote.

It can misparaphrase a real source. Sometimes the link is solid and reputable, and the synthesis still slightly overstates or twists what it says — sharpening a hedged finding into a flat assertion, or attributing a claim the source actually attributed to someone else. The link is real; the paraphrase drifted. This is the single best argument for clicking through on anything load-bearing, covered next.

It can go stale between when it ran and when you read it. On a fast-moving story, a synthesis is a snapshot. The thing it confidently reports as the latest may have been overtaken while you were reading. Freshness is the mode's strength and also its expiry date.

None of these make DeepSearch untrustworthy — they make it a *draft from a fast researcher*, which is exactly what it is. Knowing the failure shapes is what lets you read the result the right way instead of swallowing it whole.

How to Read and Verify the Sources

Here is the part people skip, and it is the part that separates research from the illusion of research. The whole value of a sourced answer is that you *can* verify it. That value is only real if you actually do, at least for the claims that matter.

You do not have to check everything. That would defeat the purpose. But you should check the load-bearing claims — the ones your decision, your argument, or your next email depends on. For those:

Click through to the actual source. A link is not proof; it is an address. Open it and confirm the source genuinely says what the synthesis claims it says. Models can misread, over-extrapolate, or attach a real link to a slightly wrong paraphrase, and the only way to catch that is to look.

Notice what kind of source it is. A primary document, a reputable outlet, a random post, and an anonymous claim are not equal weight, and a good synthesis will have told you which is which — but you should confirm it yourself for anything important. A vivid claim sourced only to unverified posts is a rumor with a link attached, not a finding.

Watch for the single-source pillar. If an entire conclusion rests on one source, that is a flag — not necessarily wrong, but fragile. Independent corroboration is what turns a report into something you can stand on.

Re-run if it is fresh. On a fast-moving story, an answer from an hour ago may already be stale. These modes are cheap to re-run relative to their value; on something genuinely live, run it again rather than trusting a result that has had time to age.

The posture to carry out of this chapter is the one this whole book keeps returning to: Grok at its best is a fast, current, capable researcher whose work you still read. DeepSearch and DeeperSearch give you the researcher; the reading is what makes the research real, and on the claims that matter it is not optional. Do that, and you have something rare — current, broad, sourced understanding in the time it takes to make coffee. Skip it, and you have only the confident *appearance* of understanding, which is worse than having none and precisely the trap Chapter 23 is about.

Chapter 10: Live Knowledge from X

A product I'd been waiting on shipped while I was asleep. By the time I sat down with coffee, the official announcement was a few hours old, the early reviews were trickling out, and — this was the part I actually cared about — the people who had stayed up to try it had spent the night posting about what worked and what didn't. I didn't want the press release; I'd read that. I wanted the verdict from the people in the trenches. So I asked Grok: "What are people actually saying about it now that they've had hands on it overnight?" And it told me — the consensus enthusiasm, the two complaints that kept recurring, the one feature everybody was surprised by, and the loud minority who hated a change, with examples I could click through to. In ninety seconds I had the read I'd have spent an hour assembling by hand.

That is the X-native advantage, and it is the most structurally distinctive thing about Grok. The previous chapter was about turning real-time access into formal research. This chapter is about the rawer, more immediate side of the same edge: asking what is happening *right now*, in the live human conversation, and reading the pulse of it. It is also where I owe you the most honest warning in the book, because the same firehose that makes Grok uniquely current makes it uniquely easy to be confidently misled. Both halves of that are true at once, and using Grok well means holding both.

The Structural Advantage

Most AI assistants can search the web. That is real, and it is useful. But the web is, in a sense, the *settled* layer of the internet — articles get written, pages get published, and there is a lag between a thing happening and a thing being written up. X is the other layer: the live, unsettled, in-the-moment reaction, where an enormous share of the world's response to events shows up first, in human words, as it happens. A goal scored, a product shipped, a statement made, a market moving — the first wave of human reaction lands on X before it lands anywhere a traditional search would find it.

Grok was built by the company that owns X, and it has native, first-class access to that firehose. This is not a bolted-on integration that occasionally checks a social feed; it is a built-in line to the live stream of what people are posting and discussing. That is the capability competitors do not match in the same way. They can search the web after it settles. Grok can read the conversation while it is still happening.

What that unlocks, concretely:

"What's happening right now." Ask about a live event — an unfolding situation, an ongoing announcement, something that started an hour ago — and Grok can read the live posts and tell you the current state, not the state as of some training cutoff.

Live sentiment and reaction. Not just *what* happened but how people are *responding* — the mood, the split between the cheering and the groaning, the take that is catching fire, the backlash building under the official line. This is genuinely hard to get any other way, and it is often the thing you actually want.

Breaking developments. On a fast-moving story, the firehose updates faster than the write-ups. Grok can surface what is new in the last little while, which is exactly the window where settled sources have nothing yet.

Understanding threads and accounts. Point Grok at a long thread, an exchange, or an account's recent posting, and it can summarize the argument, explain the context, and tell you what is being said without you scrolling through all of it yourself.

There is a fourth use that does not get named as often but quietly earns its keep: **catching the early signal on a topic you track but cannot watch.** You follow a niche — a sport, a company, a technology, a neighborhood issue — closely enough to care but not closely enough to refresh a feed all day. The settled layer will tell you what mattered, eventually, in a write-up. The live layer tells you it is happening while it is happening. Asking Grok "anything new today on [topic]?" once or twice a day is a cheap way to stay current on something without surrendering your attention to a feed, and because Grok is reading the live conversation rather than a curated news section, it surfaces the small-but-real developments that never become an article. That is a different value than breaking-news drama; it is the steady, low-grade currency that the firehose makes possible and the settled layer, by its nature, lags on.

Two Layers, Side by Side

It helps to make the two-layer picture concrete, because the whole skill of this chapter is knowing which layer you are reading. Suppose a mid-sized company announces a recall at 9 a.m. The settled layer — the layer a normal web search reaches — fills in over hours: the company's own press page goes up, a wire service writes it, a trade outlet adds context by lunchtime, a careful analysis appears the next morning. Each of those has passed through some hands. The live layer, meanwhile, is already roaring by 9:05: the customer who noticed the defect first, the employee hinting at what really happened, the competitor's fans dunking, the lawyer fishing for plaintiffs, the three accounts spreading a wrong version of the story faster than the right one. Both layers describe the same event. They are not interchangeable. The live layer is faster and richer and far less trustworthy; the settled layer is slower and thinner and far more reliable. Grok can read both, and the reason it is worth learning to ask precisely is that an undirected question ("tell me about the recall") gives you a blend you cannot un-blend, while a directed one ("what's the live reaction on X" versus "what's the confirmed reporting") keeps them apart where you can weigh each on its own terms.

Asking About the Present

The skill here is partly knowing what kinds of questions this capability is *for*. With a frozen assistant you learn not to ask about anything recent, the way you stop expecting a printed encyclopedia to cover this morning. With Grok's live edge, a whole class of question becomes worth asking:

"What's the reaction to the thing that was announced this morning?" "What are people saying about this right now?" "Is there a developing story about X today?" "What's the live sentiment around this — are people happy or angry?" "Summarize what's being said in this thread." "What's the mood on this topic this afternoon versus last week?"

These are questions about the *present tense of human opinion*, and they were essentially unanswerable by an AI assistant until real-time access existed. The phrasing that works best names the time and the kind of signal you want: not just "what about this product" but "what's the reaction to this product now that it's out." You are telling Grok to go read the live conversation, not to recite what it already knew.

A practical note: live results are timestamped reality, which means they age. An answer about live sentiment is a snapshot of a particular moment. If the thing is still moving, the answer is perishable in the most literal sense — re-ask in an hour and it may have shifted. That is a feature, not a bug; it is what current means.

There is also a craft to the *grain* of the time window you ask for, and it is worth being deliberate about it. "What are people saying about this?" with no time anchor invites Grok to blend the last week with the last ten minutes, and the blend is usually less useful than either alone. "What's the reaction in the last hour?" pins it to the freshest signal — right for a story that is actively moving. "How has the mood shifted over the past week?" asks for a trajectory rather than a snapshot — right when you want to know whether something is building or fading, not just where it stands right now. Those are genuinely different questions, and naming the window tells Grok which one you mean. A useful habit: when a topic is hot, ask narrow ("last hour," "this afternoon"); when you want to understand a trend, ask wide ("this week versus last week," "over the past month"). The same live edge serves both, but only if you say which you want.

One more bit of phrasing that pays off: ask for the *shape* of the conversation, not just its content. "Is the reaction mostly positive, mostly negative, or split?" and "what are the main camps, and what is each one saying?" pull back a structured read instead of an undifferentiated wash of takes. The live conversation almost always has a structure — a loud faction, a quieter majority, a contrarian corner — and asking Grok to surface that structure is far more informative than asking it to summarize "the reaction" as though there were one. When you want to actually understand a moment rather than just sample it, ask for the divisions.

A Worked Read: Following a Story as It Breaks

Let me walk one all the way through, because the difference between a vague question and a useful one only shows up in the doing. Say a streaming service you pay for goes down on a weekday evening. Your own app just spins. The first instinct is to wonder whether it is you — your wifi, your account, your device — or them. That is a perfect live-edge question, and the phrasing matters.

The weak version is "is [service] down?" Grok can answer it, but you have given it nothing about time, so it may reach for stale signal. The strong version names the moment and the signal you want: "Is [service] having an outage right now? What are people reporting in the last half hour?" Now Grok reads the live conversation and comes back with the texture you actually need — yes, a spike of outage reports starting around 8:40, concentrated in a couple of regions, the service's own status account acknowledging "elevated errors," and a scattering of people saying it just came back for them. In one read you have learned three things that change your behavior: it is them, not you (stop rebooting your router); it is acknowledged (someone is on it); and it may already be recovering (wait ten minutes before doing anything drastic).

Then follow up, because the first read is rarely the whole story. "Has the company said anything about cause or timeline?" surfaces the official post if there is one. "Is this the same outage as the one last week or a new one?" pulls the recent pattern. Each follow-up rides on the same live thread, and each is a present-tense question a frozen assistant simply cannot field. Notice what you did *not* do: you did not ask Grok to *promise* the service is down, because "lots of people are reporting an outage" is a strong live signal but still a report. In this case the report is cheap to confirm — try the service yourself in ten minutes — so the stakes of being wrong are near zero. Hold that thought, because the next section is about the cases where they are not.

It is worth contrasting that low-stakes read with a higher-stakes one to see how the same capability changes character. Swap the streaming outage for something with money or safety attached — say you hold a position in a stock and you see a flicker of something on your feed suggesting the company just disclosed bad news. The live question is identical in shape: "Is there a developing story about [company] in the last hour — any reports of an announcement, a disclosure, a leadership change?" And Grok will read the live conversation and hand you the texture: a cluster of posts citing a filing, a few accounts claiming insider knowledge, a lot of speculation, maybe the official account silent so far. The mechanics are the same as the outage. What changes is everything downstream. With the outage, "probably down, try again in ten minutes" is a fine place to stop, because being wrong costs nothing. With the stock, "probably bad news, lots of people saying so" is precisely the place you must *not* stop, because acting on an unconfirmed live claim with real money is how people get hurt — both by being wrong and by being early to a rumor that someone else planted on purpose. Same tool, same query pattern, opposite discipline. The live edge tells you *something is happening and here is what people say it is*; whether that is the end of your work or merely the beginning depends entirely on what it costs to be wrong. Learning to feel that distinction in the moment — to notice when a cheap-to-verify lead has quietly become an expensive-to-act-on one — is most of using the live edge like an adult rather than a gambler.

Asking About Accounts and Threads

A second mode of live use is narrower and underrated: pointing Grok at a *specific* slice of the conversation rather than the whole topic. A long, contentious thread has spun out forty replies deep and you want the gist without living through every quote-tweet. An account you do not follow has been posting all week about a thing you care about, and you want the throughline, not the scroll. A debate between two people has gotten heated and you cannot tell who said what first. These are summarization-over-live-text tasks, and Grok is good at them: "Summarize the argument in this thread and tell me where the disagreement actually is." "What has this account been saying about [topic] over the past week?" "Walk me through how this exchange started and how it escalated."

The value here is compression. The live layer is enormous and most of it is noise even when it is honest noise — repetition, dunks, side-quarrels. Asking Grok to read a bounded slice and hand you the structure of it turns an hour of doomscrolling into a paragraph. The same caution carries, though: a summary of a thread is a faithful account of *what the thread says*, and a thread can be wrong in chorus. Compression does not add verification; it just makes the unverified material easier to read.

There is a particular flavor of this that is worth singling out, because it is where account-level summarization is most useful and most prone to a subtle distortion. Suppose you need to get up to speed on a public dispute — two researchers, or two companies, or two commentators, who have been arguing in public for a week, and you have walked in halfway. Scrolling the raw exchange is brutal: it is non-linear, full of quote-tweets and replies-to-replies, and the order things appear in your feed has nothing to do with the order they were said. This is exactly the kind of bounded, structured slice Grok is good at untangling. "Reconstruct this disagreement in order: who said what, when, and what's the actual point of contention underneath the noise?" can turn a week of tangled posting into a clean narrative you can actually follow. But notice the distortion risk: any summary of an argument quietly decides whose framing is the spine of the story, and a model summarizing a heated exchange can absorb the framing of whichever side posted more or louder. So the useful follow-up is adversarial: "Now give me the strongest version of each side's case, steelmanned, not as they insulted each other." That forces a balanced read out of material that was anything but balanced, and it is far more useful for actually understanding a dispute than a tidy summary that has unconsciously picked a winner.

The Honest Part: Live Information Is Noisy

Now the warning, and I am going to give it the weight it deserves, because everything good about this chapter comes with it attached.

Real-time access to the live conversation is a superpower *and a new way to be wrong*. The web that settles into articles has, for all its faults, been through at least some filter — an editor, a fact-check, a delay during which the obviously false often gets corrected. The live firehose has not. What lives on X, alongside the genuine eyewitness reports and the sharp first reactions, is rumor, unverified claims, coordinated propaganda, jokes mistaken for sincerity, confident nonsense, and outright lies. During a breaking event this is at its worst: the early hours of a developing story are precisely when the most false information circulates, because nobody has confirmed anything yet and everybody is posting anyway.

This means a few things you have to internalize:

"People are saying it" is not "it is true." Grok can accurately report that a claim is circulating widely and the claim can still be completely false. High volume is not verification; sometimes it is the opposite, because falsehoods that are exciting spread faster than truths that are dull. A summary of live sentiment is a faithful read of *what is being said*, which is a different thing from *what is so*.

Sentiment can be manufactured. The mood on a topic can be amplified by coordinated activity, by bots, by a motivated faction posting hard. "Everyone is furious about this" may mean a loud, organized few are furious. Live sentiment is real signal, but it is signal about the conversation, which is not always a representative sample of reality or even of opinion.

Breaking-event noise is the danger zone. The very situations where the live edge is most thrilling — the unfolding story, the developing crisis — are the situations where the live stream is least reliable. The earliest "facts" of a breaking event are routinely wrong. Treat anything urgent and unconfirmed as a lead to verify, never as established.

To make that danger concrete: in the first chaotic hour of almost any major breaking event, the live conversation reliably produces a few predictable kinds of falsehood. There is the misidentified subject — a name attached to the wrong person, spread by thousands before anyone checks. There is the recycled image — a dramatic photo from a different event years ago, presented as happening now, often with no malice, just someone reposting in the heat of it. There is the premature death or premature all-clear — a claim that resolves the uncertainty everyone is anxious about, which is exactly why it spreads whether or not it is true. And there is the deliberate plant — the account seeding a false narrative because the confusion is the point. Grok, reading honestly, will report what is circulating. If you ask it "what happened" during that first hour, it can hand you a clean, fluent summary of claims that later turn out to be wrong — not because Grok failed, but because the conversation it read was wrong, and a faithful summary of wrong inputs is wrong. The fix is not to distrust Grok; it is to know which hour you are in. The first hour of a breaking story is the hour to read for *leads* and trust nothing; the picture firms up as the settled layer catches up.

Using It Responsibly

None of that means don't use the live edge. It means use it the way a good journalist uses a tip: as a starting point that points you toward the truth, not as the truth itself. The discipline is straightforward and it is worth making a habit:

Separate report from fact in your own head. When Grok tells you what people are saying, hold it as *reportedly* until you have reason to upgrade it. The phrasing of a good answer will often help — it will say "users are reporting" or "unconfirmed claims suggest" — and you should preserve that hedge rather than rounding it up to certainty.

Check the sources on anything that matters. Live findings come with the posts behind them; click through. Notice whether a striking claim traces to a credible firsthand account or to an anonymous post with no corroboration. The same source-reading discipline from Chapter 9 applies here, doubly, because the source quality on the live layer is wildly uneven.

Triangulate against the settled layer for facts. Use the live conversation to learn what is happening and how people feel; use traditional, verifiable sources to confirm what is actually *true*. When a live claim is important, a DeepSearch across both layers, or a check against a primary source, is how you turn a fast lead into something you can rely on. Chapter 11 builds exactly that combined workflow.

Be most skeptical when it's most exciting. The claim that is spreading fastest and feels most consequential is the one to slow down on. That instinct — to verify hardest precisely when verification feels least necessary — is the whole game.

The live knowledge from X is the thing Grok has that the others don't, and on the days you need to know what is happening *now*, it is genuinely without substitute. But it hands you the conversation, raw, and the conversation contains everything humans say, true and false alike. The deeper failure modes get their own treatment in Chapter 23, and what happens to your own X data sits in Chapter 25. For now the rule is the one I'd give a new reporter on a fast story: the firehose will get you there first, and it will also get you there wrong if you trust it blind — so read it eagerly, and believe it carefully.

Chapter 11: Research Workflows

A friend who runs a small business called me, rattled. A supplier she depended on had gone quiet — missed shipments, an unanswered email, and then a vague post online that something was "being restructured." She needed to know, fast, whether she was about to lose a supplier and should be lining up a replacement, or whether this was a non-event she could ignore. That is a research question, and it is exactly the kind that lives in the seam between the live conversation and the settled record. So we worked it together, and the way we worked it is this chapter: not a single magic query, but a sequence — frame, search, triangulate, confirm — that combines DeepSearch, the live X edge, and Think mode into something you can actually trust enough to act on.

The previous two chapters gave you the instruments. DeepSearch and DeeperSearch are the sourced-research engine; live knowledge from X is the real-time pulse; Think mode (Chapter 8) is the careful reasoning you reach for when the analysis is hard. This chapter is about playing them together. A good research workflow is mostly about asking the right thing in the right order and knowing, at each step, whether you need the live layer, the settled layer, or your own judgment.

Frame the Question Before You Search

The most common research mistake is searching before you have decided what you are actually trying to learn. "Tell me about this supplier" is a weak prompt because it does not say what *decision* the answer serves. "Is there credible evidence that this supplier is in financial trouble or shutting down operations, in the last few weeks?" is a strong one, because it names the claim to test and the time window that matters.

Framing well means deciding three things up front:

What you actually need to know — the specific question whose answer changes what you do. Not the topic, the question. My friend did not need a history of the supplier; she needed "should I line up a replacement now."

How current it has to be — whether this is a settled-record question (use the web and authoritative sources) or a happening-now question (the live X edge earns its keep) or, as is common, both.

What "good enough to act" looks like — the bar for proof. A casual curiosity needs less; a decision with money on it needs corroboration before you move.

Get those three set and the rest of the workflow mostly follows. Skip them and you will get a fluent answer to a question you did not quite mean to ask.

Worked Example: The Wobbly Supplier

Here is how the supplier question actually went, because the sequence is the lesson.

First, the live read. We started on the X edge, because the only signal so far was a vague online post and we wanted the immediate conversation. The question was framed for the present: "What are people saying right now about [supplier] — any reports of financial trouble, layoffs, or closure in the last couple of weeks?" Grok surfaced a small cluster of posts — a couple from people claiming to be former employees mentioning layoffs, some customer complaints about missed orders, and the company's own carefully worded "restructuring" post. Useful — but every bit of it was *reportedly*, exactly the kind of unverified live claim Chapter 10 warns about. We had leads, not facts.

Second, the deep search. Now we escalated to DeepSearch to pull the settled layer and synthesize across both. The question got sharper: "Find any reliable reporting, filings, official statements, or news about [supplier]'s financial condition or restructuring in the past month, and tell me what's confirmed versus rumored." DeepSearch came back with a real trade-press item confirming the layoffs, the company's official statement (which confirmed restructuring but denied closure), and a note that the missed-shipment complaints were widespread enough to appear across multiple independent sources. Crucially, it kept the confirmed and the rumored separate — the layoffs and restructuring were established; the "they're shutting down entirely" claim traced only back to anonymous posts and was not corroborated anywhere credible.

Third, the reasoning. The facts were now in hand but the *judgment* — what should she do — was still open, and it was genuinely a multi-factor call: confirmed instability but not confirmed collapse, real fulfillment problems, an official denial of the worst case. This is where Think mode earns its place. We asked Grok to reason through it deliberately: given a supplier with confirmed layoffs and restructuring, ongoing fulfillment failures, but no credible evidence of imminent closure, what is the risk-weighted move for a small business dependent on them? It worked through it step by step and landed where a careful advisor would: the evidence justified *quietly lining up an alternative now* without yet *cutting ties*, because the downside of being caught without a backup outweighed the cost of preparing one. That was a defensible decision, built on verified facts rather than a rumor and a vibe.

The shape of that example is the reusable pattern: **live edge for the immediate signal, DeepSearch to verify and synthesize across the settled record, Think mode to reason the verified facts into a decision.** Most serious research questions decompose into some version of those three moves.

It is worth pausing on *why* this order works rather than just that it does, because the reasoning generalizes past suppliers. The live edge goes first not because it is most trustworthy — it is the least — but because it is the fastest way to learn *whether there is anything to research at all*, and what shape the question really has. The vague online post told my friend almost nothing; the first live read told her there were specific, nameable claims (layoffs, missed orders, restructuring) worth checking. That reframed the research from "is something wrong?" into "are these particular claims true?" — a far more tractable question to point DeepSearch at. The live layer, in other words, earns its place at the front of the workflow precisely *because* it is raw: it is reconnaissance, and reconnaissance is supposed to be fast and provisional. The mistake is not using it first; the mistake is stopping there.

A Second Worked Example: A Settled-Record Question

Not every research task starts on the live edge, and it is worth seeing one that mostly doesn't, because the instinct to reach for the firehose is sometimes the wrong instinct. Suppose you are weighing whether to switch your small team onto a particular software tool, and the question that actually matters is "does this tool's licensing let us use it the way we intend — internal commercial use, a handful of seats — without a surprise cost or a legal snag?" That is almost entirely a settled-record question. The answer lives in a license document, a pricing page, the vendor's terms, maybe a clarifying post from the maintainers. The live conversation will have opinions, and they will mostly be noise for this purpose — someone's hot take on the company is not the license.

So the framing flips the usual order. You start with DeepSearch pointed at authoritative material: "Find the actual license terms and the official pricing for [tool], and tell me specifically whether internal commercial use with a small number of users is permitted and what it costs." DeepSearch fetches the primary documents and synthesizes them, and here the discipline is to make it show its work — you want the clause, not a paraphrase, because licensing is exactly the domain where a confident summary can quietly drop the qualifier that matters. If the answer hinges on a term of art (and licensing answers often do), this is a good moment for Think mode: "Given this license text, walk through whether our specific use — five internal users, a commercial product, no redistribution — is clearly permitted, clearly prohibited, or genuinely ambiguous." Reasoning slowly over the actual text catches the ambiguity a fast answer papers over.

Where does the live edge come in at all? At the margins, as a check: "Have there been recent reports of this vendor changing their licensing or pricing, or of users hitting unexpected costs?" That is a present-tense question — license terms and prices drift, and a change announced last week may not be in the cached documents yet. So even a settled-record task often ends with one live query, used precisely for the thing the live layer is good at: catching the recent change the settled record hasn't absorbed. The lesson is the mirror image of the supplier example: there, the live edge led and the settled record confirmed; here, the settled record leads and the live edge catches drift. Knowing which order a question wants is most of the craft.

Following Up and Triangulating

Research is rarely one query. The follow-up is where the real understanding gets built, and Grok holds the thread across a conversation so you can drill in without re-explaining from scratch.

The follow-ups that pay off most:

"What's the source for that specific claim?" when something matters or surprises you. Pull the individual assertion out and make it justify itself.

"Is there anything that contradicts this?" to actively hunt for the other side rather than letting an early, tidy answer settle into a conclusion. A research answer that has never been challenged is one you have not finished checking.

"What's confirmed versus still rumored?" to keep the live-layer leads and the settled-layer facts from blurring together in your head as the conversation gets long.

"What would change this conclusion?" to surface the assumptions the answer rests on, so you know which new fact would flip it.

Triangulation is the spine of all of this: a claim you can trace to multiple independent sources is something you can stand on; a claim resting on one source — especially one live post — is a lead you are still trusting on faith. When the stakes are real, you are not done until the important claims are corroborated, and "Grok said so confidently" is not corroboration. Grok is the researcher assembling the material; the triangulating judgment stays yours.

A subtle trap inside triangulation is worth naming, because it catches careful people: *independent* sources, not merely *multiple* sources. Three articles that all trace back to the same original post are not three confirmations; they are one claim wearing three coats. The live layer is especially prone to this — a single unverified post gets screenshotted, quote-tweeted, and written up by a fast aggregator, and within an hour the same assertion appears in a dozen places that look distinct but share one ancestor. When a claim matters, the question to push on is not "how many sources" but "do these sources actually know independently, or are they echoing one another?" A useful follow-up to ask Grok directly: "Do these sources corroborate each other independently, or do they all trace back to the same origin?" Sometimes it can tell you, and the answer reframes how much weight the cluster deserves.

It also pays to ask Grok to argue against its own finding. "Make the strongest case that this conclusion is wrong" is one of the most productive prompts in research, because a model that has assembled a tidy answer will, if asked, surface the counter-evidence it glided past on the way there. You are not looking to be talked out of the conclusion; you are looking to see whether it survives contact with its best objection. A finding that holds up after you have honestly tried to break it is worth far more than one that merely sounded confident the first time.

A related habit catches a quieter failure: the gap between what the sources actually said and what the summary implies. Models, like people, smooth. A source that said "the company is exploring options" can quietly become "the company is in trouble" by the time it reaches a tidy summary, and the verb did most of the damage. So when a conclusion rests on a particular claim, it is worth pulling the claim back to the ground: "Quote the exact passage that supports this, and tell me whether the source states it or merely suggests it." That distinction — states versus suggests, asserts versus speculates, confirms versus reports — is where overconfident research goes wrong, and it is invisible in a fluent paraphrase. Asking for the literal language is how you check that the summary has not promoted a hedge into a fact on its way to you. It is slower. It is also the difference between knowing something and having merely been told it persuasively.

Knowing When Live Helps and When It Doesn't

A real research skill is knowing which layer a question even belongs to, because reaching for the wrong one wastes effort or, worse, imports noise where you needed rigor.

Live data helps when currency is the whole point: breaking developments, present-tense sentiment, reactions to something recent, fast-moving situations where the settled record hasn't caught up yet. Here the X edge is close to irreplaceable, and a frozen assistant would simply fail.

Authoritative sources are what you need when the question is about what is *established* rather than what is *current*: how a system works, what a law says, what a primary document actually states, settled facts that won't change by this afternoon. Pulling these from live posts is a way to get a confident, popular, wrong answer. For these, you want the document, the reputable outlet, the primary source — and DeepSearch is good at fetching exactly those when you point it that way.

Most real questions are mixed, like the supplier: a live component (what's the immediate signal) and a settled component (what's actually confirmed). The workflow handles that by using each layer for what it is good at — live for the pulse, authoritative for the proof — and never letting the first stand in for the second. The failure mode to avoid is letting the excitement of the live finding substitute for the verification that turns it into a fact. Slowing down to confirm exactly when the live answer feels most compelling is the discipline the whole of Part III has been building toward.

Assembling Findings You Can Trust

When you have done the work, the last step is putting it into a form you and others can rely on, and Grok helps here too. Ask it to assemble the verified findings into a brief — what is confirmed, what is still uncertain, what the sources are, and what follows from it — and you get a clean artifact instead of a scattered chat history. But the trustworthiness of that brief is a property of the process behind it, not the polish on top of it. A confident summary built on unverified leads is more dangerous than no summary, because it *looks* finished.

So the closing discipline is the same one this part keeps returning to, stated as a checklist you can carry into any research task: framed the real question, used the live edge for currency and the settled record for facts, separated confirmed from rumored, traced the load-bearing claims to independent sources, and reasoned the verified material — not the raw material — into the conclusion. Do those, and Grok turns from a fast answer machine into something rarer and more valuable: a research partner whose work is current, sourced, and sound enough to act on. The instruments are powerful. The trust comes from how you play them.

Part IV: Creating with Grok

Audience: everyone. Grok doesn't only read and reason — it makes things: images, video, spoken conversation, and an eye for what you show it.

"A tool that can both see what you give it and make what you ask for is no longer a chat box. It's a studio with a search engine attached."

— A working principle

Chapter 12: Aurora: Image Generation

I needed a picture of a lighthouse for a talk — not a stock photo, but a specific one: a squat stone lighthouse on a rocky coast at dusk, fog rolling in low, one lit window. I typed roughly that sentence into Grok and asked for an image. A few seconds later I had it, and the thing that stopped me was not that it existed but how *photographic* it looked — the haze on the rocks, the warm window light bleeding into the gray, the wet sheen on the stone. It did not look illustrated. It looked taken. That is Aurora, and the photorealism is the headline.

This chapter is the first of Part IV, where we leave comprehension behind and start *creating* with Grok. Aurora is xAI's own image-generation model — text-to-image — and it is the engine behind every picture Grok makes for you. This chapter is about what it is, how to ask it for what you actually want, what it is genuinely good and genuinely bad at, and the responsible-use line you need to hold from the very first image. That last part is brief here on purpose; Chapter 26 is where the ethics get the full treatment they deserve.

What Aurora Is

Aurora is xAI's text-to-image model. You describe a picture in words; it generates the picture. It is not a separate app you go to and it is not a third-party tool bolted onto Grok — it is xAI's own model, built in-house, and it is what answers when you ask Grok to "make an image of," "draw," "show me," or "generate a picture of" something. You are already in the right place; the image capability lives inside the same conversation as everything else.

The trait that defines Aurora, and the reason it earned attention when it landed, is **photorealism**. Plenty of image models render in a recognizable "AI art" style — glossy, slightly plastic, vaguely the same no matter the prompt. Aurora is notably strong at producing images that read as *photographs*: real lighting, real texture, real depth of field, the small imperfections that make an eye accept an image as a captured moment rather than a rendered one. It does stylized and illustrated work too, but the photographic register is where it stands out, and it is worth reaching for that strength deliberately.

One honest caution up front, the one this whole book keeps making: **Aurora is the current image engine as I write, and the specifics drift.** The model name, the exact look, the resolution, the limits on how many images you can make and how fast — all of that moves, and some of it varies by tier (Chapter 5). Verify the particulars against xAI's current documentation. The durable thing is the capability: Grok can make photorealistic images from a sentence, and using that well is a skill you can learn now and keep.

How to Prompt Aurora Well

The difference between a disappointing image and a striking one is almost always the prompt. A bare noun — "a dog" — gives the model nothing to work with and it will guess at everything: breed, setting, lighting, mood, framing. The fix is not to write a longer prompt for its own sake; it is to *decide* the things you care about and leave the rest open. A few habits do most of the work.

Name the subject, then the scene around it. "A border collie" is a subject. "A border collie mid-leap catching a frisbee on a beach at sunset" is a subject in a moment. The model renders moments far better than it renders nouns, because a moment implies lighting, posture, and setting all at once.

Say what kind of image it is. This is the single highest-leverage word in most prompts. "A photograph of," "an oil painting of," "a watercolor of," "a 3D render of," "a pencil sketch of" — the medium steers the entire look. If you want Aurora's photorealism, *ask for a photograph*, and Aurora will usually reward you for it.

Describe the light. Photographers obsess over light because it makes the image, and Aurora responds to it. "Golden hour," "harsh midday sun," "soft overcast light," "a single lamp in a dark room," "backlit, rim-lit" — naming the light is often the difference between a flat picture and one with depth and mood.

Specify framing and distance when it matters. "Close-up," "wide shot," "overhead," "from below," "portrait orientation" — if the composition matters to you, say it, or the model will pick for you.

Iterate instead of agonizing. You almost never get the perfect image on the first try, and that is fine — generation is fast and conversation is the interface. Look at what came back, decide the *one* thing most wrong with it, and say so: "make it nighttime," "pull back, show the whole building," "warmer light," "fewer people in the background." Aurora is far easier to steer in small corrective steps than to nail in one heroic prompt. Treat the first image as a draft you are editing, not a slot machine you keep pulling.

A Worked Prompt: From Vague to Sharp

The lighthouse worked on the first try because I happened to phrase it well, but most images do not, so let me walk through one that needed steering, because the editing loop is the real skill. Say I want an image for a blog post about working from a home office, and I start where most people start: "a home office." What comes back is generic — a tidy desk, a laptop, a window, the kind of stock-looking interior you have seen a thousand times. It is not wrong; it is just nothing. The fix is to decide what I actually want and add it one dimension at a time.

First, the medium and the light, because those are highest-leverage: "a photograph of a home office, early morning, soft light coming through a window from the left." Already better — now it reads as a captured room rather than a render, and the directional light gives it depth. But it is still empty of character. So I add the moment: "a photograph of a cozy home office in the early morning, soft window light from the left, a steaming mug of coffee on a wooden desk, a sleeping cat on the windowsill." Now there is a story in the frame, and the model has concrete things to light and place. Suppose the result is close but the desk looks too corporate for the warm tone I want. One corrective step: "warmer, more lived-in — wooden desk, a few books stacked, plants." Suppose the cat came out slightly wrong, a common failure with animals. Another step: "lose the cat, keep everything else." Four short turns, each fixing exactly one thing, and I have a usable image — far faster than trying to write the perfect 60-word prompt up front and far more reliable, because at each step I am reacting to what Aurora actually did rather than guessing what it will do.

The pattern generalizes. Start broad enough to see what the model reaches for, then make a sequence of single-variable edits — medium, then light, then subject detail, then the one specific thing that is off. Resist the urge to change three things at once, because when you do and the result shifts, you cannot tell which change caused what. Small steps keep the loop legible.

One thing the editing loop teaches that no list of tips quite conveys: Aurora is far better at *adding* and *replacing* than at *removing*. "Lose the cat" sometimes leaves a cat-shaped smudge, because negation is genuinely hard for image models — telling a model "no cat" can make it think about cats. When you want something gone, it often works better to re-describe the scene without it than to ask for its removal: instead of "remove the people in the background," say "a quiet, empty home office," and let the absence be part of the new description rather than a subtraction from the old one. The same applies to unwanted style: "less cartoonish" is weaker than re-anchoring with "a realistic photograph, natural lighting." You are not so much correcting the previous image as re-aiming at a fresh target that happens to lack the thing you did not want. Once you internalize that the model thinks in terms of what to put *in* the frame rather than what to keep *out* of it, a lot of frustrating edits start landing on the first try.

What Aurora Is Good At — and What It Isn't

Let me be a guide here rather than a brochure, because knowing the failure modes saves you the most time.

It is good at: photorealistic scenes, landscapes, objects, food, architecture, atmosphere and mood, lighting, and plausible everyday imagery — the lighthouse, the dog on the beach, a product on a clean surface, a mock-up of a room. It is strong when the goal is "a believable picture of a thing or a place." It is also good at stylized and illustrated looks when you ask for them by name. For concepting, mood boards, slide imagery, social posts, placeholder art, and visual brainstorming, it is genuinely useful and genuinely fast.

It struggles with the things image models as a category still struggle with, and you should expect these rather than be surprised by them:

- **Text inside images.** Words rendered *in* a picture — a sign, a label, a book cover, a logo — frequently come out garbled, misspelled, or melted. This is improving across the field, but do not rely on Aurora to produce correct, readable text in an image. If the words matter, plan to add them yourself afterward.
- **Hands, fingers, and fine anatomy.** The classic tell. Extra fingers, fused fingers, hands in impossible positions. Better than it was, still not solved. Check hands.
- **Precise counts and exact spatial relationships.** "Exactly five birds," "the cup to the *left* of the plate, the fork on top" — the model approximates rather than counts, and exact arrangements are hit-or-miss.

- **Consistency across images.** Asking for "the same character" in a second image will usually give you a *similar* person, not the identical one. Maintaining a consistent face, outfit, or object across multiple generations is hard and unreliable.
- **Faithful real-world detail.** It will happily invent a plausible-looking map, diagram, or piece of machinery that is wrong in every specific. Aurora makes things that *look* right; it does not verify that they *are* right.

The honest summary: Aurora is excellent at producing convincing imagery and unreliable at producing *correct* imagery where correctness is countable, spellable, or factual. Use it for the former; do not trust it for the latter without checking.

Working Around the Weak Spots

Knowing the failure modes is half the value; the other half is the small set of workarounds that turn each one from a dead end into a manageable annoyance.

For text in images, stop asking Aurora to render the words at all. Generate the image with empty space where the text should go — "a poster with a large blank banner across the top" — and add the real, correct text afterward in any image editor or slide tool. You get Aurora's photorealism for the picture and crisp, accurate type for the words, instead of the melted compromise you get when you ask for both at once.

For hands and anatomy, two moves help. First, compose them out of the frame when you can — a portrait cropped at the chest dodges the hand problem entirely, and a person holding a mug hides the fingers that would otherwise betray the model. Second, when hands must be visible, generate a few variations and pick the least-broken one; the failure is probabilistic, so more pulls give you a better best-of. And always check hands before you use an image of people, because they are the detail a viewer's eye snags on.

For counts and arrangement, lower your expectations and verify by looking. If you genuinely need exactly three objects in a specific layout, Aurora is the wrong tool to enforce that; generate something close and arrange the precise version yourself, or accept "a few" instead of "exactly three." Asking the model to count is asking it to do the one thing it does not really do.

For consistency across images — the same character, the same product, twice — accept up front that this is the hard, unreliable case. You can improve your odds by being extremely specific and repeating the exact same descriptive phrases each time, and starting from the same reference image where the tool supports it, but do not build a project that *depends* on identical recurrence across many generations. If your work needs a consistent recurring character, that is a sign the job has outgrown a general image model.

Each of these is the same underlying move: don't fight the model on the thing it is bad at — route around it, and reserve Aurora for the thing it is genuinely great at, which is making a single convincing picture look real.

Realistic Uses

Where does this actually earn its place in a week of work? In my experience, the wins are these. **Illustrating a presentation or a post** without hunting through stock libraries or paying for licenses. **Concepting and mood boards** — generating a dozen visual directions in minutes so a team can point at the one that feels right before anyone commits real design time. **Placeholder and mock-up art** for a draft, a wireframe, a pitch, where you need *something* in the frame and will replace it later. **Visual brainstorming** — seeing an idea rendered often clarifies it faster than describing it. **Personal and creative projects** where the fun is in the making.

What it is *not* a drop-in replacement for is professional photography of real things, brand-critical final art, anything requiring legal clearance, or imagery that has to be factually precise. It is a fast, capable first-draft tool for visuals, and like every other capability in this book, the output is something to use with judgment, not swallow whole.

The Responsible-Use Line (the Short Version)

I am going to be direct, because this is where Aurora's reputation comes from and where you can get yourself into real trouble. Aurora has a **looser-guardrails reputation** than some competing image models — it will attempt things others refuse. That is a capability and a liability in the same breath, and the responsibility lands on you.

Two areas demand care from the first image.

Real people and likenesses. Generating realistic images of identifiable real people — public figures or anyone else — is the fastest route to harm here: misinformation, fabricated "photographs" of events that never happened, deepfakes, and images that defame, deceive, or violate someone's likeness rights. The technical ability to make such an image is not permission to. Treat real-person generation as something you do rarely, deliberately, and never to deceive — and understand that "it looked real" is exactly the problem, not a feature.

Sensitive content. Just because a model will attempt something does not make it appropriate to make, to share, or to make at work. The looser the guardrails, the more the judgment has to come from you rather than from the tool refusing on your behalf.

This is the brief version, and brief is all it should be here. **Chapter 26 treats the full ethics of responsible use at work** — likeness, consent, disclosure, deepfakes, and the policies that govern all of it — properly and at length. For now, carry one rule out of this chapter alongside the photorealism: the more convincing the image, the more it matters what you choose to make.

That is Aurora — a fast, strikingly photorealistic image engine that lives right inside your conversation with Grok, rewards a prompt that names medium and light, fumbles text and hands, and hands you a power that is yours to wield carefully. Next we keep the pictures moving: Grok Imagine turns this same impulse — describe it, get it — toward short video.

Chapter 13: Grok Imagine: Video

Type fifteen words describing a paper boat drifting down a rain-gutter stream, ask Grok Imagine for video, and in under a minute you have four seconds of exactly that — leaves spinning past, the camera tracking alongside. It is not Pixar. The water moves a little strangely if you stare, and a leaf flickers in and out of existence. But it *moves*, it holds together, and a sentence in your head has become a moving picture on your screen. The right response to that is two things at once: genuine delight, and a quick recalibration of what to expect from a tool like this. Both reactions are correct, and this chapter is about holding them at the same time.

Grok Imagine is xAI's short-form video generation — image-and-text-to-video — and it sits right beside Aurora in Part IV as the second creative engine inside Grok. Where Aurora makes a still image from a sentence, Grok Imagine makes a brief moving clip. This chapter is about what it actually does, what to realistically expect from it, how to prompt it, and where it helps and where it does not. The single most useful thing I can give you up front is calibrated expectations, so let me start there and be honest about it.

Set Expectations First: Short, Improving Fast, Not Hollywood

I am going to lead with the limitation rather than bury it, because the fastest way to be disappointed by AI video is to walk in expecting a film and the fastest way to be delighted is to walk in knowing what it is.

Grok Imagine makes short clips. We are talking seconds, not minutes — brief moving moments, not scenes with a beginning, middle, and end. You do not type a screenplay and get a movie. You describe a single moment of motion and get a few seconds of it.

It is improving fast. AI video as a whole is moving more quickly than almost anything else in this field right now — the clips getting longer, smoother, more coherent, and more controllable release over release. Whatever its exact limits are when you read this, assume they are looser than what I am describing, because this is the most perishable capability in the book and it is moving in one direction. **Verify the current clip length, resolution, and the available controls against xAI's documentation**; treat any specific number I give as already aging.

It is not Hollywood, and the failure modes are real. Be ready for the characteristic glitches of generated video: physics that looks slightly wrong, objects that flicker or morph, faces that drift between frames, motion that warps when it gets complex, and a general inability to hold a consistent character or object across a clip the way a real camera does. These are not bugs you can prompt away entirely; they are where the technology is. The clips are genuinely impressive *for what they are* — a sentence becoming motion in seconds — and genuinely not a substitute for filmed footage. Hold both.

Get those three facts settled and everything else about Grok Imagine becomes useful rather than disappointing.

What It Actually Does

Grok Imagine takes **text, an image, or both**, and produces a short video clip.

Text-to-video is the simplest path: you describe the motion you want — "a paper boat drifting down a rain-gutter stream, camera tracking alongside" — and it generates a clip of it. You are not describing a still and hoping; you are describing *movement*, and naming the movement is the whole skill.

Image-to-video is the one that often delights people most. You give it a still image — one you made with Aurora, or a photo — and ask it to bring that image to life: the clouds drift, the hair moves in the wind, the water ripples, the camera slowly pushes in. This is a powerful pairing with Aurora from the previous chapter: generate the exact still you want, then animate it. Starting from a fixed image also tends to produce more controlled, coherent results than starting from text alone, because the model has a concrete frame to work from rather than inventing everything.

The output is a brief clip you can save and share. Like every creative output in this book, it is a draft to use with judgment, not a finished broadcast asset to trust blind.

A Worked Clip: Animating a Still

The most reliable way into Grok Imagine is the image-to-video path, so let me walk one through, because it shows both the power and the seams. Suppose I made an Aurora still I am happy with — that coastal lighthouse at dusk, fog low on the rocks, one lit window. As a photograph it is finished. But I want it to *breathe* for a short social clip, so I hand it to Grok Imagine and ask: "animate this — fog drifting slowly across the rocks, the lit window glowing steady, very slow camera push toward the lighthouse." Notice what I asked for. I did not ask for a lot of motion; I asked for a *little*, and I named the camera. Under a minute later I have four seconds where the fog rolls, the window holds its warm light, and the frame eases gently inward. It looks intentional, almost cinematic, and the reason it works is that I gave the model an easy job: a calm scene, one simple atmospheric motion, one slow camera move.

Now watch what happens when I get greedy. Suppose on the next pull I ask for "waves crashing hard against the rocks, seabirds flying past, the camera swooping around the lighthouse." Three independent motions and an aggressive camera, and the clip falls apart in the characteristic ways — the waves morph instead of crash, a bird smears into the rock and vanishes, the lighthouse subtly warps as the camera arcs because the model is inventing geometry it never saw. Same tool, same starting image, wildly different result, and the only variable that changed was how much I asked it to coordinate at once. That contrast is the single most useful thing to internalize about AI video today: coherence is a budget, and complexity spends it fast. The calm clip looked good because it stayed within budget. The busy one looked broken because it blew past it.

So the working method is to start an animation from a still you already like, ask for the smallest motion that achieves the effect, and add ambition only if the simple version holds. Anchoring to a fixed image is doing a lot of quiet work here, because the model begins from a concrete frame rather than hallucinating the whole world — which is why image-to-video so reliably beats text-to-video for anything you actually want to look controlled.

There is a deeper lesson hiding in that contrast, and it is the one that separates people who get usable clips from people who get a folder of glitchy near-misses. The motions that survive are the ones that *do not require the model to invent geometry it cannot see*. Fog drifting is safe because fog has no fixed shape to violate — it can morph all it likes and still look like fog. A slow push-in is safe because moving straight toward a scene reveals almost no new surfaces; the model is mostly scaling what it already has. Contrast that with a crashing wave, which has to generate plausible new shape every single frame, or an orbiting camera, which demands the back of a building the model never saw and must fabricate. Once you think in those terms, you can predict in advance which prompts will hold and which will fall apart, and you start *designing* clips around motions that are cheap to render coherently: atmosphere over hard mechanics, gentle camera over sweeping camera, one moving element over many. The art of AI video right now is mostly the art of choosing motions that do not stress the seams.

How to Prompt It

Prompting video is prompting Aurora plus *motion*, and the motion is where people under-specify.

Describe the movement, not just the scene. A still prompt names what is in the frame; a video prompt also names what *happens*. "A city street at night" is a picture. "A city street at night, neon signs flickering, steam rising from a grate, a single car passing through frame" gives the model things to animate. If you only describe a static scene, the model has to invent the motion, and invented motion is where the weirdness creeps in.

Name the camera. This is the high-leverage trick. "Slow push in," "camera pans left," "tracking shot following the subject," "static locked-off shot," "drone shot pulling up and back" — camera language does enormous work, because controlled camera motion reads as intentional and cinematic, while uncontrolled motion reads as glitchy. A slow, simple camera move over a mostly still scene is one of the most reliable ways to get a clip that looks good.

Keep the motion simple. Complexity is where coherence breaks. One subject doing one clear thing, or a calm camera move over a quiet scene, holds together far better than a crowded frame full of independently moving parts. Ambitious, busy prompts are where you get the morphing and flickering. If you want a clean clip, ask for a simple moment.

Start from an image when you can. As above — anchoring the clip to an Aurora still or a real photo gives you more control over how it looks before anything starts moving, and tends to reduce the drift you get from pure text.

Iterate. Same posture as Aurora. Generate, watch, adjust one thing — "slower camera," "calmer water," "make it daytime" — and go again. Fast generation means cheap iteration; use it.

Reading the Glitches: What Each One Tells You

When a clip comes back wrong, the *way* it is wrong usually tells you what to change, so it is worth learning to read the failures rather than just re-rolling blindly. A few of the common ones and their tells.

Morphing and melting — an object's shape sliding around between frames, a face subtly rearranging — almost always means the scene asked for too much motion or too much detail to track. The fix is to simplify: fewer moving elements, a calmer camera, a clearer single subject. Morphing is the model telling you it could not hold everything together at once.

Flickering objects — a leaf, a bird, a small prop popping in and out of existence — usually means a detail too small or too numerous for the model to commit to. Drop the fussy small elements, or make the one you care about larger and more central so the model treats it as load-bearing.

Warping geometry on camera moves — a building bending as the camera arcs around it — comes from asking the camera to reveal angles the model has to invent. The fix is the gentlest camera that gets the job done: a slow push or a small pan reveals little new geometry and warps less than a sweeping orbit.

Drift in faces and characters — a person's features shifting across the clip — is the consistency problem from the previous chapter, now playing out over time instead of across separate images. Keep clips short, keep the face partly turned or in motion that hides the drift, and do not build anything that needs a recognizable person to stay exactly themselves for several seconds.

The throughline is the same budget idea: nearly every glitch is the model running out of room to stay coherent. When you see one, the productive response is usually to *ask for less* — less motion, less detail, less camera — rather than to add corrective instructions on top of an over-ambitious prompt.

Realistic Uses

So what is a few seconds of slightly imperfect motion actually *for*? More than you would guess, as long as you match the tool to jobs that suit short clips.

Social and short-form content is the natural home — clips built for platforms where short *is* the format and a few seconds of motion is exactly the unit. **Animating a still you already love** — bringing an Aurora image or a favorite photo to gentle life. **Concepting motion** — showing a client, a team, or yourself how an idea might *move* before committing real production budget, the way Aurora concepts a look. **Quick visual ideas, mood pieces, and creative play**, where the fun and the value are in the making and the looseness is part of the charm rather than a flaw.

There is also a workflow worth naming that gets more out of short clips than their length suggests: treat each generation as a *shot*, not a film. A few seconds is roughly the length of a single cut in fast-paced editing, and a handful of separately generated clips — a slow push on the lighthouse, a detail of the lit window, a wide of the fog-bound coast — can be assembled in any ordinary video editor into something that reads as a deliberate little sequence, with music and a cut hiding the seams. You are not asking one prompt to produce a coherent thirty-second scene, which it cannot do; you are generating thirty-second's worth of usable two-second moments and editing them together yourself. This is how the limitation stops being a wall and becomes a constraint you work within — the same way a photographer shoots many frames to keep a few. The model makes shots; you make the cut.

What it is *not*, today, is a replacement for real video production: not for long-form footage, not for anything needing consistent characters across multiple shots, not for precise, reliable, broadcast-grade output. It is a fast, playful, improving-by-the-month tool for *short* motion, and it is best used where short, generated, and good-enough-and-getting-better is genuinely what the job calls for.

What to Carry Forward

Grok Imagine is the moving-picture sibling of Aurora: describe a moment of motion — or animate a still — and get a few seconds of video in seconds of waiting. The discipline that makes it work is mostly expectation-setting. Lead with the truth that it makes *short* clips, that it is *improving fast* enough that today's limits are tomorrow's footnotes, and that it is decidedly *not Hollywood* — the glitches are part of where the technology sits. Prompt the *movement*, name the *camera*, keep it *simple*, anchor to an *image* when you can, and iterate cheaply. And note, before we move on, that everything from Chapter 12's responsible-use line applies here with more force, not less: a convincing fake still image is one problem, and a convincing fake moving one is a larger version of the same problem. Chapter 26 carries that weight.

With Aurora and Grok Imagine, you have seen Grok create — pictures and motion from a sentence. The rest of Part IV turns the lens around: instead of Grok making media for you, Grok takes media *in* — your voice, and the images and documents you show it.

Chapter 14: Voice Mode

I worked out the shape of a difficult email while walking the dog. Not at my desk, not typing — walking, leash in one hand, phone in my pocket, talking out loud to Grok through an earbud. "I need to tell a vendor we're not renewing, but keep the door open, and not sound cold about it. Help me think through how to say it." It asked me a clarifying question. I answered while crossing the street. By the time I was home I had the three sentences I needed, and I had never looked at a screen. That is voice mode, and the thing it changed for me was not *what* Grok could do but *when* and *where* I could reach for it.

This chapter closes the human-facing inputs of Part IV. Voice mode lets you speak to Grok and hear it answer — a spoken, hands-free, conversational way to use the same assistant you have been typing to. This chapter is about how it works, what it is genuinely good for, where it falls short, and one pairing that makes it more than a novelty: voice plus Grok's real-time knowledge. Let me start with what it actually is, because the mental model matters.

How It Works

Voice mode is a spoken conversation with Grok. You talk; it listens, understands, and answers out loud in a synthesized voice; you talk back. There is no command syntax to memorize and no wake-word ritual to get right — once you are in voice mode it is closer to a phone call than to barking orders at a smart speaker. You speak naturally, in whole rambling sentences if you like, and it follows.

Under the surface, three things are happening: your speech is transcribed to text, the same Grok model you already know reasons over that text and composes a reply, and the reply is spoken back to you in a natural-sounding voice. The important implication of that middle step is that **the intelligence is the same Grok.** Voice is an *interface*, not a lesser model — the reasoning, the knowledge, the ability to search live information, all of it is the assistant you have met in the rest of this book. You are not talking to a stripped-down voice assistant; you are talking to Grok, out loud.

As always, treat the surrounding specifics as perishable. The available voices, exactly where voice mode shows up, how you start it, and which tiers include it all drift and may vary by platform (Chapter 5 for tiers, Chapter 6 for surfaces). **Confirm the current details against xAI's documentation.** The durable capability is simple and stable: you can have a real spoken back-and-forth with Grok.

What It's Genuinely Good For

Voice is not better than typing at everything — it is better at *specific* things, and worse at others, and the trick is knowing which is which. Here is where it genuinely wins.

Hands-free, eyes-free moments. The whole point. Driving, walking, cooking, exercising, carrying things, standing in a line — any moment where your hands or eyes are busy but your mind is free. The dog-walk email is the canonical case: I could not have typed that, but I could absolutely talk it through. Voice turns dead time into usable time, and that is its largest practical value.

Thinking out loud. This one surprised me with how useful it is. There is a real cognitive difference between *typing* a thought and *speaking* it. Speaking is looser, faster, more associative — you ramble toward an idea instead of composing it. For brainstorming, for working through a decision, for talking yourself to the center of a half-formed thought, having a competent listener that talks back is genuinely clarifying. You explain the problem aloud, it reflects and questions, and somewhere in the back-and-forth the answer surfaces. Some thinking happens better in the mouth than in the fingers.

Practicing conversations. A quietly excellent use. Rehearse the hard talk before you have it — the negotiation, the interview, the feedback you need to deliver, the pitch, the difficult conversation with a colleague. Speaking out loud to something that responds is far closer to the real thing than writing a script you will never read aloud, and it surfaces the places you stumble while the stakes are zero. Language practice belongs here too: speaking a language you are learning, out loud, with something patient on the other end.

Quick questions on the go. The small stuff — a fact, a conversion, a reminder of a name, a "what's the deal with X" — answered while your hands are full and your screen is away.

Accessibility and reading load. This one is easy to overlook if typing has never been a barrier for you, but for some people voice is not a convenience, it is the difference between using the tool and not. If a screen is hard to read for long, if typing is painful or slow, if a motor or vision difference makes the keyboard a wall, then a spoken back-and-forth is not the lesser interface — it is the primary one. And even for people with no such barrier, voice lowers the cost of *starting*: there is a real activation energy to opening an app and composing a typed question that simply is not there when you can ask out loud. A surprising number of questions get asked, and answered, that would never have been worth the friction of typing. Lowering the threshold to ask is its own kind of value.

A Worked Session: Rehearsing the Hard Conversation

The rehearsal use deserves a full walk-through, because it is the one people underrate most and it is genuinely transformative once you try it. Say I have to give a direct report critical feedback tomorrow — their work has slipped and I need to say so without crushing them, and I am dreading it. Reading a script will not prepare me, because the hard part is not the words, it is saying them out loud and handling what comes back. So I rehearse with voice mode.

I start by framing it as a role-play: "I need to give an employee feedback that their recent work has been below standard. Play the employee — react the way a defensive but not unreasonable person might — and let's run through the conversation. I'll open." Then I actually say my opening line, out loud, the way I would tomorrow. Grok answers in character, maybe a little defensive, maybe with an excuse, and now I have to respond to a real reaction instead of a blank page. I stumble on the second exchange — I hear myself over-apologizing, softening the message until it disappears — and *that is the value*, because I caught it in a rehearsal where the stakes are zero instead of in the room where they are not. I stop, step out of the role-play — "let's pause; how did my framing land, and where did I undercut myself?" — and Grok shifts from playing the employee to coaching me. Then I run it again, tighter. By the third pass I have found the three sentences that carry the message cleanly, and more importantly I have *said them with my mouth*, so tomorrow they will not be the first time.

That loop — role-play, stumble, step out and reflect, run it again — works for the negotiation, the interview, the pitch, the apology, the boundary you need to set. Voice makes it real in a way text cannot, because the channel you are practicing in is the channel you will perform in. And because it is just Grok underneath, you can switch fluidly between *being in* the scene and *analyzing* the scene, which is exactly what a good rehearsal coach does.

A small refinement makes the rehearsal sharper: tell Grok how hard to push. Left to its own defaults, a role-played counterpart tends toward the reasonable middle — a little defensive, but cooperative. Real conversations are not always so accommodating, and the value of a rehearsal is partly in handling the version that goes badly. So set the difficulty explicitly: "play the employee as genuinely upset and a bit unfair — make me work for it," or, on the next pass, "now play them as cold and minimal, the kind who just says 'fine' and shuts down." Running the hard conversation against several temperaments — the defensive one, the wounded one, the stonewaller, the one who turns it back on you — surfaces different stumbles, and you walk in tomorrow having already met the worst-case version once. The point of rehearsal is not to script the conversation you hope to have; it is to be unsurprised by the one you might actually get.

Where It Falls Short

An honest guide names the limits, and voice has clear ones. Knowing them keeps you from being annoyed when voice is the wrong tool.

Anything visual or precise is awkward. Voice is a terrible way to receive a long list, a table of numbers, code, a URL, a spelling, or anything you need to *see* and *keep*. Hearing a paragraph read aloud is fine; hearing twelve bullet points read aloud is forgettable noise. When the answer is something you need on screen — to copy, to scan, to reference — typing and reading beat voice every time. Many people do their best work by speaking the question on the go and then *reading* the considered answer later on a screen.

Spoken misunderstandings happen. Transcription is good, not perfect. Names, technical terms, homophones, and accented or noisy speech all introduce errors, and an error in what Grok *heard* propagates into the answer. If a reply seems to be answering a slightly different question than you asked, suspect a mishearing before you suspect the model.

It is not built for length or for keeping. A long, layered answer is hard to absorb by ear and harder to retain. Voice shines for short exchanges and conversational thinking, not for delivering a research report you will need to consult again.

It can be socially and contextually wrong. Talking to your phone is fine on a walk and out of place in a quiet office, a meeting, or a crowded train. Voice has a *where*, and it is not everywhere.

Interruption and turn-taking are imperfect. A spoken conversation depends on knowing when the other party is done, and a voice assistant can misjudge it — cutting in while you are mid-thought, or talking past the moment you wanted to jump in. The smoother voice gets, the less this bites, but it is still a real friction, and it is most annoying in exactly the thinking-out-loud use where you want to ramble without being interrupted. If you find the back-and-forth choppy, a small adaptation helps: speak in slightly more complete chunks and pause cleanly at the end, which gives the system a clearer signal that your turn is over.

The spoken voice can flatten nuance. A subtler limit than the others, and worth naming because it shapes what voice is good for. A synthesized voice reading an answer aloud tends to smooth the emphasis that punctuation and formatting carry on a screen — the difference between a confident claim and a hedged one, the weight of a "but," the qualifier buried in a clause. On the page your eye catches "this is *probably* true, though one source disputes it"; by ear, in a steady synthesized cadence, the "probably" and the dispute can slide past. So for anything where the *certainty* of the answer matters as much as its content — a medical question, a legal nuance, a factual claim you intend to act on — the spoken version can leave you more confident than the words actually warrant. This is a quieter cousin of the read-it-later pattern below: when the hedges matter, see the answer, do not just hear it.

The honest frame: voice is a superb interface for *some* moments and a poor one for others. Reach for it when your hands or eyes are occupied or when speaking helps you think; reach for the keyboard when you need to see, keep, or precisely control the result.

The Read-It-Later Pattern

There is a hybrid that resolves most of voice's weaknesses, and once you find it you will use it constantly: **ask by voice on the go, read the considered**

answer later on a screen. The two failure modes that hurt most — that voice is bad at anything you need to *keep*, and bad at anything long or precise — both dissolve if the spoken exchange is the *capture* step and the screen is the *consume* step.

In practice this looks like talking through a question while you walk or drive — "help me outline the three sections of this proposal, I'll flesh it out later" — letting the conversation do the thinking part, and then opening the same thread on your phone or laptop afterward to read, copy, and refine the result with your eyes. You got voice's real gift, which is using the time when your hands were busy, without paying voice's real tax, which is that spoken output evaporates. The spoken channel is where you *think and capture*; the screen is where you *keep and use.* Treating them as two halves of one workflow, rather than competing interfaces, is how heavy voice users actually get the most out of it.

Pairing Voice with Real-Time Questions

Here is where voice stops being a convenience and becomes something the other assistants struggle to match. Remember that voice is just an interface over the *full* Grok — and the full Grok has live access to X and the web (the heart of this book, and the subject of Part III). Put those together and you can ask, out loud, hands-free, about *what is happening right now.*

"What's the latest on the storm heading our way?" while you are packing the car. "What are people saying about the game?" while you are walking to it. "Did that announcement actually happen yet, and how's it landing?" while your hands are full. These are not memory-bank questions a frozen voice assistant could answer — they are *present-tense* questions, and Grok can go look and tell you, spoken, in the moment you most want to be hands-free. The on-the-go situations where voice is most useful are exactly the situations where being *current* matters most, and Grok is one of the few assistants where the voice interface sits on top of genuine real-time knowledge. That pairing — spoken, hands-free, *and* current — is voice mode's quiet superpower, and it is worth reaching for deliberately.

What to Carry Forward

Voice mode is the same Grok, spoken — an interface, not a lesser model — and the intelligence, the reasoning, and the live knowledge all come along. It is genuinely good for the hands-free and eyes-free moments that fill a day, for thinking out loud, for rehearsing the conversations that matter, and for quick questions on the move. It is genuinely poor at anything visual, precise, lengthy, or meant to be kept, and it can mishear, so a reply that seems slightly off is often a mistranscription rather than a mistake in reasoning. And its best trick is the pairing this whole book keeps circling back to: a spoken, hands-free question about *right now*, answered from live sources, in the exact moment your hands are full.

Chapter 15: Vision: Images and Documents

Picture an error screen you have never seen before — a wall of red text, a stack trace, a configuration value highlighted in yellow. Instead of typing it out, you take a screenshot, drop it into Grok, and ask, "What's failing here and what should I check first?" It reads the screen — the actual pixels, the highlighted line, the path in the trace — and names the likely culprit and the three things to look at, in order. Say it gets two of the three right. The point is that you never transcribed a word. You showed it the problem the way you would show a colleague leaning over your shoulder, and it *saw* it.

This is the last chapter of Part IV, and it closes the loop. Aurora and Grok Imagine were Grok *making* media; voice was Grok *hearing* you. Vision is Grok *seeing* — you give it a photo, a screenshot, a chart, or a document, and it describes, extracts, and reasons about what is in it. This chapter is about what that actually buys you, the practical uses worth building habits around, and — critically — where vision errs and why you have to verify what it reads. That last part is not a footnote; it is the discipline that makes vision trustworthy, so I will not let you leave the chapter without it.

What Vision Is

Vision is Grok seeing an image you provide and reasoning about it. You attach a picture — a photo, a screenshot, a scanned page, a chart, a diagram, a document — and ask a question about it, and Grok answers based on what it actually observes in the image. This is the inbound half of multimodality: Chapter 12 was Grok generating pixels; this is Grok *reading* them.

It is worth being concrete about how much "reasoning about an image" covers, because it is more than "describe the picture." Grok can:

Describe a scene. Tell you what is in a photo — the objects, the setting, the people, what is happening, the mood.

Read text in an image (OCR). Pull the words out of a screenshot, a photographed sign, a scanned page, a receipt, a whiteboard, a handwritten note. This is one of the most useful everyday capabilities — text trapped in a picture becomes text you can work with.

Read charts and graphs. Look at a bar chart, a line graph, a pie chart, and tell you what it shows — the trend, the comparison, the outlier, the takeaway — rather than just noting that a chart exists.

Answer questions about a document. Hand it a page, a form, a contract excerpt, a menu, a spec sheet, and ask specific questions — "what's the deadline in this," "which plan includes X," "summarize this section," "what does clause 4 actually require" — and it reads and answers.

Reason across what it sees and what it knows. The error-screen story is this: it did not just transcribe the stack trace, it *interpreted* it, combining what it read with what it knows about how such errors arise. Vision is not a scanner; it is the full Grok, looking.

The last item is the one people underestimate, so it is worth slowing down on. A scanner gives you back the pixels as text and stops; a vision model gives you back the pixels as *meaning*. Show it a photo of a parking sign with four lines of conflicting restrictions and ask "can I park here at 6pm on a Tuesday," and it does not merely read the sign — it reads the sign, parses the time windows, and answers the actual question. Show it a screenshot of two side-by-side phone plans and ask "which is cheaper if I use about 20 gigabytes a month," and it reads both tables, does the arithmetic, and gives you a recommendation. The image is the input; the reasoning is the product. That is the difference between vision as a feature and vision as part of an assistant.

As ever, treat the edges as perishable — which surfaces accept images, how many at once, what file types and sizes, and whether your tier includes it all drift (Chapters 5 and 6). **Confirm against xAI's current documentation.** The durable capability is stable: show Grok a picture, ask about it, get a reasoned answer.

How to Ask: Showing Grok an Image Well

The quality of what you get back from vision is shaped as much by how you ask as by what you show, and a few habits raise the hit rate noticeably.

Ask a specific question, not "what is this." "Describe this image" gets you a generic description; "what's the total at the bottom of this receipt, and what's the tax line separately" gets you the two numbers you actually need. Vision rewards a pointed question the same way a human helper does — tell it what you are looking for and it looks for that.

Give it the cleanest image you can. This is not vanity; it is accuracy. A photo taken straight-on in decent light, in focus, with the relevant part filling the frame, will be read far more reliably than the same content shot at an angle in shadow. If you are photographing a document to extract numbers from it, the thirty seconds spent flattening the page and getting the light right buys you a materially lower error rate. Crop to the part that matters when you can; a tight crop on the one clause you care about reads better than the whole crowded page.

Tell it the context it cannot see. Grok reads the image, not the world around it. "This is a screenshot from our internal billing tool" or "this chart is from a competitor's earnings deck" gives it the frame to interpret what it is looking at, and the interpretation improves accordingly. The picture shows the *what*; you often have to supply the *what it is*.

Chain follow-ups in the same conversation. Once Grok has the image in context, you can keep asking — "now compare that to the figure in the second image," "rewrite the deadline clause in plain English," "which of these line items is the subscription." The image stays loaded; you interrogate it. Vision is at its best as a conversation about a picture, not a single round-trip.

Ask it what it is unsure about. This one is underused and pays for itself. After a read of anything dense — a form, a table, a low-quality photo — add "and tell me which parts you were least confident reading." Grok will often flag the smudged figure, the field it had to infer, the column it was not certain it aligned correctly. That is a free map of exactly where to point your own eyes during verification. It does not always know what it got wrong — a confident misread by definition does not feel uncertain to the model — but when it *can* tell you the read was shaky, that is the cheapest verification lead you will get.

Practical Uses

This is one of the most quietly useful things Grok does, and the uses are everywhere once you start noticing them.

Troubleshooting from a screenshot. The error message, the confusing settings page, the app that is misbehaving, the spreadsheet formula throwing a fault — screenshot it and ask. Far faster than transcribing, and Grok sees the context around the problem, not just the words you would have typed.

Pulling text out of images. A photographed document, a receipt for an expense report, a whiteboard after a meeting, a slide from a deck, a sign, a business card — capture it and have Grok extract the text into something you can edit, search, or paste. This alone earns vision a place in a normal week.

Understanding charts and visuals. A chart in a report, a graph in an article, an infographic, a diagram you do not follow — show it and ask what it means, what the trend is, or what the catch is. Useful when the visual is doing the talking and you want it in words.

Working through documents. A contract clause, a form you are not sure how to fill in, a dense spec, a menu in a language you do not read, a page of instructions — point Grok at it and ask the specific question you have. It reads the page and answers from it.

Identifying and explaining what you are looking at. A plant, a part, a landmark, a UI you do not recognize, an unfamiliar object — "what is this and how do I use it" — answered from the picture.

The everyday "what am I looking at." Sometimes you just want a fast read on something visual, and showing beats describing.

Real-time, applied to a picture. Here is a use that only a tool wired into the present can do: you photograph something you cannot identify — a product on a shelf, a logo, a piece of news on someone else's screen — and ask not just "what is this" but "what is this, and what are people saying about it right now." Vision names the thing; the real-time search (Part III) tells you the live context around it. The picture is the query you could not type because you did not have the words for it; the present is the answer you could not get from a frozen reference. The two capabilities composing is a small thing that feels surprisingly futuristic the first time it works.

The thread through all of these is the same as the error screen: **showing is faster than telling**, and a great deal of the information we deal with lives in images and documents rather than in clean text. Vision is the bridge from that visual world into everything else Grok can do with it.

Three Walkthroughs

The categories above are tidy; real use is messier, so here are three fuller examples of vision doing actual work, end to end.

The receipt pile. You come back from a conference with a fistful of receipts to expense, and the company portal wants merchant, date, amount, and category for each. Instead of typing twelve receipts, you photograph them and hand Grok one at a time: "Pull out the merchant name, the date, the total, and tell me if this looks like a meal, transport, or lodging." For most receipts it reads all four fields cleanly and you paste them straight into the form. For one — a crumpled cab receipt with the total half-faded — it gives you a number, and this is exactly where the chapter's later warning lives: you glance at the image, see the total is genuinely ambiguous, and read it yourself. The win is real (eleven of twelve done in a fraction of the time) and the discipline is exactly where it needs to be (you checked the one that mattered). That mix — fast on the clean cases, verified on the hard one — is what good vision use feels like in practice.

The chart in the article. You are reading a piece that hinges on a chart you cannot quite parse — two lines crossing, a shaded region, an axis in some unit you are not sure about. You screenshot it and ask, "What's this chart actually showing, and what's the takeaway the author wants me to draw from it?" Grok describes the two series, identifies the crossover point, and tells you the implied argument. Useful — but now do the thing the chapter insists on: if you intend to *quote* a specific value off that chart ("revenue crossed costs in Q3"), confirm that reading against the chart yourself, because a transposed axis or a misjudged crossover is precisely the kind of confident error vision makes. Use it to *understand* the chart; verify any number you will repeat.

The form in another language. You are handed a one-page form in a language you do not read — a customs declaration, a rental agreement, a medical intake sheet. You photograph it and ask Grok to translate it field by field and tell you what each blank is asking for. It walks you through the form, flags the two fields that are mandatory, and explains the one piece of fine print at the bottom. This is vision plus translation plus reasoning composed into a single answer, and it turns a genuinely stuck moment into a five-minute task. The honest edge: for anything legally binding, this is orientation, not authority — you understand the form well enough to act, but a signature on a contract you only understood through a vision translation still deserves a human who reads the language.

The whiteboard after the meeting. A working session ends and the wall is covered in boxes, arrows, half-sentences, and three different people's handwriting. The honest reality is that within a day none of you will reliably remember what the diagram in the corner meant. So before anyone wipes it, you photograph the whole board and ask Grok to "transcribe everything on this whiteboard, group it by the cluster it's drawn in, and turn the action items into a checklist with an owner where a name is written next to a task." What comes back is a structured artifact instead of a photo that rots in your camera roll. Two honest wrinkles show up here and they are instructive. First, handwriting is the hard case — the messier the scrawl, the more you should expect a misread, and "transcribe a colleague's cursive" is exactly where vision guesses. Second, the *grouping* is interpretation, not transcription: Grok is inferring which arrow connects to which box, and it can connect them wrong. So this is a fast, genuinely useful first pass that you then reconcile against the photo while the meeting is still fresh in your head — which is precisely when reconciling is easy and precisely when you would otherwise do nothing.

Vision Plus the Rest of Grok

The examples above already hint at it, but it is worth stating directly: vision is rarely the whole job. It is the *intake* step, and its value multiplies when what it reads flows into everything else Grok does.

Read a chart, then ask Grok to write the paragraph of analysis that goes under it. Extract the action items from a photo of a whiteboard, then ask it to draft the follow-up email to the team. Pull the numbers off a screenshot of last quarter's dashboard, then ask it to reason about what changed. Translate a foreign menu, then ask which dishes are vegetarian. In each case vision turns something visual into text, and then the assistant you already know takes over. The most useful vision habit is not "describe this image" — it is "read this, and then do the next thing with what you read." That composition is where vision stops being a party trick and becomes part of the workflow.

There is a subtle discipline hiding in this composition, and it is worth naming because it is easy to lose. When vision is the *intake* step, any error it makes does not stay contained — it propagates into everything downstream. If Grok misreads a figure off your dashboard screenshot and you immediately ask it to "reason about what changed," the reasoning is now built on a wrong number, and the analysis will be confidently, coherently wrong in a way that hides the original misread entirely. By the time the error reaches the email you send, it has been laundered through two steps and looks nothing like an OCR slip. The fix is to verify at the *seam* — confirm the extracted facts the moment vision produces them, before you chain the next request, rather than trying to catch a bad number after it has been reasoned over and dressed in prose. Check the read, then build on it. Building first and checking last is how a single misread digit ends up in something you sent under your name.

Where Vision Errs — and Why You Verify

Now the discipline, and I am going to be firm about it, because vision's failures have a particular and dangerous quality: **they are confident and specific.** Grok will read a number off a document, or a value off a chart, in a clear declarative sentence — and sometimes the number is wrong. It does not hedge the misread; it states it the same way it states a correct read. That is exactly what makes vision errors hazardous, and exactly why verification is not optional.

Here is where it most reliably slips.

Misreading text, especially numbers and the hard cases. OCR is good, not perfect. Handwriting, low resolution, blur, glare, unusual fonts, dense tables, and small text all raise the error rate — and the single most consequential failure is **getting a number wrong.** A misread digit in a price, a date, a measurement, an account number, or a total looks identical in the output to a correct one. *Always* re-check critical numbers against the source image yourself. This is the one I will repeat: do not act on a number vision extracted without confirming it with your own eyes.

Misreading charts. It can read a trend backwards, transpose two values, misjudge a proportion, or confidently report a data point that the chart does not actually support. Use it to *understand* a chart; verify any specific value you intend to use.

Confusing detail in busy images. A cluttered scene, a crowded form, overlapping elements, fine print — the more packed the image, the more likely something gets missed, merged, or invented. Dense documents are where careful checking matters most.

Inventing plausible detail. As with everything generative, vision can fill a gap with something that fits rather than something that is there — a field it could not quite read becomes a confident guess. If it matters, confirm it is *read*, not *guessed*.

Hard-to-see content. Anything partially obscured, cut off, at a steep angle, or in poor light is a guess dressed as an observation.

To make the danger concrete rather than abstract: imagine a photographed invoice where the total reads "$1,450.00" but a fold in the paper crosses the comma, and Grok returns "$1,450.00" on one read and "$1.450,00" — a European format, a wildly different number — if the lighting shifts. Or a handwritten "7" that it reads as a "1," turning a quantity of seven units into one. Or a stacked bar chart where it reports the *top* segment's value as the whole bar's total. None of these announce themselves. The output sentence looks identical whether the read was right or wrong, which is the entire problem: the error has no tell. A correct extraction and a confident misread are typographically indistinguishable, so you cannot catch the misread by reading Grok's answer — you can only catch it by looking back at the source. That is why the verification step is not a nicety you do when you have time; it is the only mechanism that exists for separating the right reads from the wrong ones.

The rule that covers all of it is the rule that covers this entire book, sharpened for vision: **use it to read and reason, but verify anything that matters against the source.** Vision is a tremendous accelerator for getting information out of images — and the source image is always right there, so checking the load-bearing facts against it costs you seconds. Spend the seconds. A misread number that you trusted is worse than no number at all, because it carries the false confidence of having been "read."

What to Carry Forward

Vision is Grok seeing — the inbound complement to the images it generates and the spoken conversation it holds. Show it a photo, a screenshot, a chart, or a document and it will describe, extract text from, read the chart in, answer questions about, and reason over what is actually there. The practical wins are everywhere once you look: troubleshooting from screenshots, pulling text out of pictures, decoding charts, working through documents, identifying what is in front of you. And the discipline is non-negotiable: vision errs *confidently*, most dangerously on numbers and on busy or low-quality images, so verify anything load-bearing against the source that is sitting right there in front of you.

That closes Part IV and the creative-and-perceptual core of Grok — making images and video, speaking and listening, and seeing. Part V turns to the work most people actually open an assistant for: writing, analysis, coding, and the ordinary productivity that fills a day.

Part V: Grok for Work

Audience: working professionals. The capability is real; so is the discipline. Everything here ships under your name, not Grok's.

"The assistant drafts, analyzes, and codes in seconds. The judgment about whether it got it right is still the job — and it's yours."

— A working principle

Chapter 16: Writing and Communication

Writing is the thing most people open an AI assistant for, and Grok is very good at it. Hand it a thorny email in three messy sentences — say, a polite but firm note pushing back on a deadline that has quietly slipped twice — tell it the relationship matters and you want to sound calm rather than annoyed, and the draft that comes back is often good enough to send after changing two words. That is the everyday reality of writing with Grok, and it is worth being honest about both halves of it: the draft is genuinely good, and you still read every line before it goes out under your name.

This chapter is about using Grok for the writing that fills a working life — emails, documents, summaries, the constant translation of half-formed thoughts into something another person can read. It is the most common professional use of any assistant, Grok included, and Grok is very good at it. It is also the use where the gap between "looks finished" and "is correct" is easiest to miss, because polished prose *reads* as trustworthy whether or not it is.

What Grok Is Genuinely Good At

Set aside the real-time story for a moment — we will bring it back, because it matters here more than people expect — and look at Grok purely as a writing tool. On the core writing tasks it is a strong frontier model, fully competitive with the assistants you may already use (Chapter 3), and the things it does well are the things you do all day.

Drafting from a sketch. The hardest part of most writing is the blank page, and this is where an assistant earns its keep fastest. Give Grok the gist — who it is for, what it needs to accomplish, the few facts that must appear — and it will produce a complete first draft in seconds. A draft you will edit is worth far more than a blank document you are staring at, and getting to "something to react to" is the single biggest time saving Grok offers a writer.

Editing and tightening. Paste in your own prose and ask it to cut the length by a third, fix the flow, remove the throat-clearing, or make the argument land harder, and it is a capable editor. I lean on this more than on drafting, honestly — my own first drafts plus Grok's editing pass tend to beat Grok's draft plus my editing, because the voice stays mine and the machine just sharpens it.

Shifting tone and register. This is one of the most quietly useful things any assistant does. The same content can be a terse Slack message, a warm note to a colleague, a formal memo to a regulator, or a plain-language explanation for a customer, and Grok moves between those registers fluently. "Make this friendlier," "make this more direct," "make this sound like a human wrote it and not a policy document" — all of these work, and they save the real labor of writing, which is rarely the words and usually the *judgment about how to sound.*

Summarizing and reshaping. Long thread into three bullets, dense report into an executive paragraph, meeting notes into action items, a rambling document into a clean structure — Grok handles all of it. Summarization is where it feels most like leverage, because reading and condensing is exactly the tedious-but-necessary work worth handing off.

Adapting one thing into many. A single piece of content rarely has a single life. The announcement you wrote needs to become a short email to the team, a one-line Slack post, a longer note for the people who were not in the room, and maybe a customer-facing version that strips the internal detail. Asking Grok to "take this and give me the Slack version, the email version, and the two-sentence version" is one of the genuinely time-saving moves it does, because the second through fourth versions are exactly the derivative labor you resent doing by hand. The thinking is in the first version; the rest is reformatting, and reformatting is what the tool is for. The caution rides along: each derived version still needs a read, because compression is where Grok is most likely to drop the one caveat that mattered or sharpen a tentative point into a flat assertion.

Briefing Grok So the Draft Is Actually Usable

The difference between a draft you change two words on and a draft you rewrite from scratch is almost always the brief you gave it. Grok is a strong writer, but it is writing toward whatever target you described, and a vague target produces vague prose. A few minutes spent on the ask repays itself many times over.

The brief that consistently works has four parts: **who it is for, what it needs to accomplish, the facts that must appear, and how it should sound.**

Compare two ways of asking for the same email. The thin version is "write an email declining the vendor's proposal." What comes back is competent and generic — it could be from anyone, to anyone. The full version is "Write a short email to a vendor we have worked with for three years, declining their renewal proposal because the price went up 40% and our budget is flat. Keep the relationship warm — we may come back next year. Don't apologize more than once, and don't explain our budget in detail." That version comes back nearly ready to send, because every choice the writer would otherwise guess at — the warmth, the brevity, the one fact to include and the one to withhold — you supplied. The labor of writing is mostly the judgment about how to sound and what to say; hand Grok that judgment explicitly and it executes it.

Show it a sample of your voice when voice matters. If you want the draft to sound like *you*, paste two or three things you have actually written and say "match this register." Grok is good at imitating a sample — far better than at inferring your voice from a description like "professional but casual." For anything that goes out under your name regularly, keeping a short paragraph of your own writing handy to prepend is one of the highest-leverage habits in this chapter.

Iterate in the conversation rather than re-prompting from scratch. The first draft is a starting position. "Tighter." "Warmer." "Lose the second paragraph." "That opening is too stiff — make it sound like a person." Each round costs a sentence and moves the draft closer. Treating writing as a short back-and-forth, rather than a single shot you accept or reject, is how the tool actually earns its keep.

Say what you do not want, not just what you do. Negative constraints steer harder than positive ones, because they head off the defaults you would otherwise spend rounds undoing. "No bullet points." "Don't open with 'I hope this finds you well.'" "Don't restate the question back to me." "No corporate filler — if a sentence could appear in any company's email, cut it." Grok, like every assistant, has house tendencies — a fondness for tidy lists, a reflex toward throat-clearing openers, a pull toward hedged neutrality — and naming the ones you dislike up front saves the iteration you would spend stripping them out. The brief that says what to avoid is doing half the editor's job before the draft exists.

A Briefing Walkthrough: The Apology That Has To Land

Worked detail makes the four-part brief concrete, so here is a harder case than the vendor decline: you have to apologize to a client for a mistake your team made, and the relationship is worth real money. The thin ask — "write an apology email for the missed delivery" — produces something that is technically an apology and emotionally hollow, the kind of note that makes a wronged client angrier because it reads as a form letter. The full brief carries the judgment: "Write a short email apologizing to a long-standing client for shipping their order three days late because of an error on our side. Own it plainly — no blaming the carrier, no passive voice hiding who messed up. State the one concrete thing we're doing so it doesn't happen again. Don't grovel, don't offer a discount (I'll decide that separately), and keep it to four sentences. Warm but not falsely cheerful." Notice how much of that is *judgment a colleague would otherwise have to supply*: the decision not to blame the carrier, the boundary around the discount, the length, the register. Hand Grok those decisions and the draft comes back close to sendable. Withhold them and you get a competent template that you then have to fix in exactly the places that matter — which is slower than briefing well the first time, and produces a worse note besides.

The Directness Trait — Asset and Risk

Grok has a distinctive voice. By design it is more willing than its more cautious rivals to be blunt, to take a clear position, to write with a bit of edge rather than corporate neutrality (Chapter 3 sketches this; Chapter 24 treats it in full). For writing, that is a genuine asset more often than you might assume. A lot of professional prose is bad precisely because it is over-hedged — buried in qualifiers, allergic to a plain statement, written to avoid offense rather than to communicate. Grok's willingness to just *say the thing* can cut straight through that, and asking it for "a direct version, no hedging" often produces the clearest draft.

But the same trait is a risk you carry into every send, and it matters more at work than anywhere else. Grok's default register can land sharper than you intend — a note that reads as refreshingly candid to you may read as curt or even hostile to the colleague receiving it, and you do not get to add the tone-of-voice footnote after it arrives. The fix is simple and it is yours to apply: read the draft as its actual recipient, not as its author. When the stakes or the relationship are real, I explicitly steer — "professional and warm," "firm but not cold" — rather than trusting the default, and I read the result with the specific person in mind. The directness is a tool. Like any sharp tool, it does what you point it at, including damage.

A small example shows both edges at once. Ask Grok to "tell a teammate their section of the report missed the deadline and held up the whole project," and the unsteered draft may open with something flatly accurate and entirely too cold — the kind of sentence that is correct and also starts a feud. The same request with "keep it collegial; I want them to fix it, not feel attacked" produces a note that still says the true thing plainly but frames it as a shared problem to solve. Neither draft is wrong about the facts; the difference is entirely in the register, and the register is the part you have to own. The lesson is not "Grok is harsh" — it is that Grok will be exactly as direct as you let it be, so the steering is your job and the read-as-recipient pass is not optional when a relationship is on the line.

The Real-Time Edge in Writing

Here is where Grok does something its peers structurally cannot, and it is easy to overlook because we do not usually think of writing as a real-time task. A great deal of professional writing is *about* something current — a response to an announcement that dropped this morning, a post reacting to a development in your industry, a note to your team about a story that is unfolding, a piece of marketing that has to reference what is happening *now* to feel alive rather than canned.

An assistant frozen at a training cutoff cannot help you write about this afternoon; it does not know this afternoon happened. Grok can. Because it reaches into live information from X and the web (Part III), you can ask it to draft a post responding to a breaking development and have it actually know what the development *is*, what people are saying about it, and what the mood around it is — and write into that context rather than around it. For anyone whose communication touches current events, live discourse, or the reaction to something recent, this collapses a two-step process — go find out what is happening, then write about it — into one. It is the real-time advantage applied to the most ordinary task in the book.

The same caution from Part III rides along: live sources are noisy, and the "current context" Grok writes from is only as good as what it pulled. If your timely email asserts a fact about the unfolding story, that fact needs the same verification any live claim does (Chapter 23). Writing about the present does not exempt you from checking the present.

To see how much this changes a real task, take a concrete one. Your industry's biggest competitor announced something this morning and your boss wants a short internal note explaining what happened and what it means for you, sent before lunch. With a frozen assistant the writing is the *easy* part — the hard part is the half hour of reading you have to do first to know what to write. With Grok you can compress both: "DeepSearch what [competitor] announced this morning and the early reaction to it, then draft a 150-word internal note for my team explaining what happened and the one thing it means for us." Grok pulls the live picture and writes into it in one motion. You still read the note — and you still verify the load-bearing claim, because if the early reaction was wrong about what was actually announced, your note inherits that error. But the two-step slog has become a one-step draft you refine, and the time saved is exactly the reading you no longer do by hand.

The same shape applies to the marketing post that has to reference a trending moment to feel alive, the customer update about an outage that is still unfolding, the comment on a development your readers already saw this morning. In all of them the value is the same: Grok knows what "now" is, so the writing can be *about* now rather than carefully vague to avoid being wrong about it.

Where Grok Is a Weaker Writer

Honesty cuts both ways, and a chapter that only listed strengths would mislead you. There are kinds of writing where Grok helps less, and knowing them keeps you from reaching for it where it costs you.

Anything that lives or dies on a voice only you have. A condolence note, a toast, a deeply personal apology, a piece whose whole value is that it sounds unmistakably like *you* — Grok can produce a competent version, but competent is the problem. The generic-but-fine register that makes it great for routine email is exactly wrong for writing whose point is to be particular. Use it to break the blank page here if you must, but expect to rewrite most of it, and never send a personal note in Grok's default voice.

Writing that depends on facts only you hold. When the substance is internal knowledge — what was actually said in a meeting Grok was not in, the real reason behind a decision, the history of a relationship — Grok can only write a plausible version of what it guessed. It will fill the gaps confidently, and the gaps are precisely the load-bearing parts. The more the writing turns on private context, the more you should give Grok the facts explicitly or write that part yourself.

Long, structured documents where consistency matters end to end. Grok is excellent at the paragraph and very good at the page; across a long document it can drift — a term defined one way early and used another way later, a claim in the introduction the body never delivers, a tone that wanders. For anything long, treat it as a drafting and section-editing tool, and keep the architectural consistency as your own job. A practical way to work within this limit rather than against it: draft long documents section by section with you holding the outline, rather than asking for the whole thing in one shot. You decide the structure and the through-line — the part that requires holding the entire piece in your head — and let Grok draft each section against your spec. The architecture stays yours, where it belongs, and the per-section drafting is where the tool is genuinely strong.

Anything where the format is rigid and unforgiving. Legal language with terms of art that mean something exact, regulatory filings with prescribed structure, a contract clause where one wrong word changes the obligation — Grok writes fluent prose that *resembles* these, and resemblance is the trap. It will produce something that looks like a proper indemnification clause and is subtly not one. For high-stakes formal writing with real legal or compliance weight, Grok is a tool for understanding and first-drafting, not for producing the final language. The expertise that knows why a particular phrasing matters is exactly the thing the model is imitating without possessing.

Review Discipline — Why It Bites Harder Here

Every chapter in this part returns to the same honest beat, and writing is where it is most tempting to ignore, so I will be plain about it. Grok can be confidently wrong, and fluent prose is the perfect disguise for it. A made-up statistic, a misremembered name, a date that is off by a year, a quote attributed to the wrong person — all of these arrive wrapped in the same smooth, assured sentences as the true parts, and there is no visual tell. The writing does not look less certain when it is wrong.

At work this matters more, not less, for a reason worth stating directly: **the email, the memo, the report goes out under your name, and the reader holds you accountable for it, not the tool.** Nobody on the receiving end knows or cares that Grok drafted it. If it contains a wrong number, a misstated commitment, or a tone that burns a relationship, that is yours.

So the discipline is non-negotiable and it is also cheap, which is the part people forget — reading a draft before you send it takes far less time than writing it from scratch, so review costs you almost nothing relative to what Grok already saved. Three things to actually check, every time the writing carries weight:

Every fact. Names, numbers, dates, claims, commitments, anything specific and checkable. If Grok stated it as fact and you cannot personally vouch for it, verify it or cut it. Treat unsourced specifics as suspect until confirmed.

The tone, as the recipient. Re-read with the actual person in mind. Does it land the way you intend, or has the directness tipped into something that will cost you?

The substance, as you. Does it actually say what *you* mean, or merely something plausible in the neighborhood? Grok writes a competent version of what it guessed you wanted; only you know whether that is what you wanted.

That posture — fast capable draft, careful human read — is the whole game with Grok and writing. Used that way it is one of the highest-leverage tools in a working life: it does the part that is slow and leaves you the part that needs a person. The writing that goes out is faster to produce and, because you are reading rather than staring at a blank page, often better than what you would have written alone.

Chapter 17: Analysis and Data

A colleague forwarded me a forty-page quarterly report and asked, essentially, "is this as bad as I think it is?" I pasted the whole thing into Grok and asked it to pull out the three numbers that actually mattered, tell me which direction each had moved, and flag anything that looked like it was being buried in the prose. In under a minute I had a clean read of a document I had been dreading. Then I did the thing this entire chapter is going to insist on: I opened the report and checked the three numbers myself. Two were exactly right. One had been transcribed correctly but interpreted backwards — Grok had read a narrowing loss as a growing one. That is analysis with Grok in a single story: a real accelerant, and a thing you check before you trust.

This chapter is about using Grok to reason over data and documents — summarizing reports, finding the signal in something long, structuring an analysis, working through numbers. It is some of the most valuable work an assistant does for a professional, and also the work where being confidently wrong does the most damage, because a wrong number presented with confidence travels straight into a decision.

What "Analysis" Actually Means Here

It is worth being precise, because "analysis" covers a few different things and Grok is differently reliable at each.

Reading and summarizing. Hand Grok a long document — a report, a contract, a research paper, a transcript — and ask what it says, what matters in it, what the main argument is, what is buried. This is comprehension over text, and Grok is strong at it. With its large context window (Chapter 4) it can take in a great deal at once and give you back a faithful, structured summary far faster than you could read the thing yourself.

Structuring and reasoning. Give Grok a question and some information and ask it to think it through — weigh options, lay out tradeoffs, build a framework, walk a decision step by step. This is where pairing with Think mode (Chapter 8) pays off: for genuine analytical work, the explicit reasoning mode produces more careful, less hand-wavy output than a snap answer, and it lets you *see the reasoning* so you can check the logic rather than just the conclusion.

Working with actual numbers. Ask Grok to compute, compare, find a trend, or reason quantitatively over data you have given it. This is the highest-value and the most dangerous category, and it gets the most scrutiny in this chapter, because it is where a confident error is least visible and most consequential.

A Worked Example: Reading a Long Document

Abstraction is easy to nod along to and hard to act on, so here is the summarizing job done properly, step by step, because the *method* is the transferable part.

Say you have a thirty-page market research report and twenty minutes. The naive move is to paste it in and ask "summarize this," which gets you a fluent paragraph that may or may not catch what you needed. The better move is to ask in layers. First: "Give me the three main conclusions of this report and the single most important number behind each." That gets you the skeleton. Then, instead of trusting it, you go deeper where it counts: "For the second conclusion, quote the exact sentence in the report it's based on." Now you can see whether the conclusion is actually supported or whether Grok smoothed an inference into a finding. Then the pointed question only you know to ask: "Does the report say anything about [the specific thing you care about], and if so, what exactly?" — which surfaces the buried detail a generic summary would skip.

The layered approach does two things at once. It gets you a better summary, because you are steering toward what matters to you rather than accepting the model's guess at what matters in general. And it builds verification into the process instead of bolting it on afterward — by asking for the supporting sentence, you have already half-checked the claim. This is the analyst's version of the read-the-code discipline that runs through this part: do not accept the conclusion without seeing what it rests on.

A Worked Example: Reasoning Through a Decision

The structuring use deserves its own walkthrough, because it is where Think mode (Chapter 8) changes the quality of the answer most.

Suppose you are deciding whether to renew an expensive software contract or switch to a cheaper competitor. You give Grok the relevant facts — the price difference, the switching cost, the features you actually use, the risk of disruption — and ask it to reason through the decision in Think mode. What comes back is not a verdict but a *structure*: the considerations laid out, the tradeoffs weighed, the case for each option made explicitly. Because the reasoning is visible, you can audit it. You might notice it weighted switching cost too lightly, or assumed a feature mattered that you do not use, and you correct it in the next turn: "We never use the reporting module — redo the analysis without it." The value here is not that Grok decides; it is that Grok builds the scaffold you make the decision on, and visible reasoning means you can see and fix where the scaffold is wrong. A good move is to then ask it to argue the *opposite* of where you are leaning — the cheapest insurance against talking yourself into a decision you already wanted.

A Worked Example: Comparing Options Across Numbers

The third category — working with actual numbers — deserves its own walkthrough, because it is where the discipline this chapter preaches gets its sharpest test. Say you are comparing three vendor quotes, each with a different pricing structure: one charges a flat annual fee, one charges per seat, and one charges a base plus usage. You want to know which is cheapest at your expected scale of forty users and roughly the usage you project. You give Grok the three structures and your numbers and ask it to compute the annual cost of each and rank them.

What comes back will look authoritative: three totals, a ranking, probably a sentence about the break-even point where the per-seat option overtakes the flat fee. And here is the trap laid bare — *that output is three claims, not three computations*. Grok reasoned about the arithmetic in prose; it did not run a spreadsheet. So the method is to never accept the ranking on its face. First, "show me each total as a worked calculation — the per-unit figure, the count, the multiplication, the sum." Now the arithmetic is visible, and a transposed number or a multiplication against the wrong base shows up in a step you can read. Second, take the one total the decision actually turns on and recompute it yourself, on a calculator or in a cell — thirty seconds for the number that will sit in a recommendation. Third, sanity-check the break-even claim, because that is the kind of derived figure where a small error in the setup produces a confidently wrong crossover point. The win is real: Grok structured a comparison that would have taken you fifteen minutes to lay out by hand. But the bottom-line number became trustworthy only when something deterministic confirmed it, and that confirmation is the part you cannot delegate.

The Real-Time Dimension

This is where Grok's defining strength reaches a kind of analysis the other assistants cannot match in the same way. A frozen model can analyze the document you paste in; it cannot analyze *what is happening right now* because it does not know. Grok can do both.

Pair Grok's reasoning with its live access to X and the web (Part III, especially live knowledge in Chapter 10) and you get analysis of the present: the live sentiment around a company or a product, the shape of a reaction to an announcement as it unfolds, the trajectory of a story over the last several hours, the mood of a market this afternoon. "What are people actually saying about this launch, and is the reaction net-positive or net-negative?" is an analytical question, and it is one only a real-time tool can genuinely answer. For anyone whose work involves reading live data — monitoring, sentiment, breaking developments — this is the differentiator, and it is the thing worth reaching for Grok specifically to do.

The catch, predictably, is the catch from all of Part III: live data is noisy and unverified. Sentiment pulled from X is a real signal, but it is a sample of a loud, unrepresentative, sometimes manipulated population, and Grok's read of it inherits every distortion in the source. Treat live analysis as a fast informed reading of a messy room, not as a measurement. It tells you something true about the conversation; it does not tell you something true about the world, and the two are easy to confuse.

It helps to be precise about what live sentiment analysis can and cannot honestly claim. If you ask "is the reaction to this product launch net-positive," a useful and defensible answer is "the loudest voices on X right now skew negative, and here are the three complaints driving it." That is a real finding — you have learned what the vocal slice is saying and why. What you have *not* learned is whether the product will sell, because the people posting about it are not the people buying it, and a launch can be a commercial success while the online conversation sulks. The distortion runs in specific, knowable directions: the angry post more than the satisfied one, the early adopter more than the eventual mainstream buyer, the coordinated campaign more than the organic reaction. Grok can read the room faithfully and still mislead you, not because it read wrong but because the room is not the world. Use the live read for what it genuinely gives you — speed, texture, the actual words people are using — and supply the grain of salt yourself.

Where Grok Errs — and How to Catch It

Here is the honest accounting, and it is sharper for data than for almost anything else Grok does. The failures cluster in a few predictable places.

Arithmetic and quantitative reasoning. Language models, Grok included, can make errors in calculation and numeric reasoning that a calculator never would — a multiplication that is off, a percentage computed against the wrong base, a total that does not sum, a trend read in the wrong direction. The model is reasoning about numbers in words, not running a spreadsheet, and the result can be wrong while looking completely authoritative. This is the single most important thing to internalize about analysis with Grok: **the numbers it gives you are claims, not computations, and they need to be checked like claims.** Where the math matters, verify it independently — recompute it, or have Grok show its work step by step (Think mode helps here) so you can follow each step rather than trusting the bottom line.

Misreading the source. As in my quarterly-report story, Grok can transcribe a figure correctly and then interpret it wrong — invert a direction, conflate two line items, attribute a number to the wrong period or the wrong entity. The fix is to trace any load-bearing number back to the document and confirm both the value *and* what it means.

Confident interpolation. Asked about something not actually in the data, Grok may fill the gap with a plausible-sounding inference rather than saying it does not know. An analysis can quietly drift from "what the document says" to "what a reasonable document might say," and the two read identically. When a conclusion seems to rest on a fact, check that the fact is actually present in what you gave it.

Spurious structure. Grok is good at producing clean, confident frameworks — three drivers, four risks, a tidy ranking — and the tidiness is seductive. Sometimes the structure reflects the data; sometimes it is imposed on data that does not really support it. A neat answer is not a correct one, and the cleanliness of an analysis is no evidence of its truth. Watch in particular for the suspiciously round framework — the "three reasons," the "four pillars," the ranking with no ties. Real data is lumpier than that. When the structure is conspicuously balanced, ask yourself whether the world is really that tidy or whether the model rounded the world into a shape that reads well. The format that flatters the reader is often the format that has quietly smoothed away the inconvenient parts.

False precision. A close cousin of spurious structure: Grok will report a figure as "$43,287" or "a 17.4% increase" when the data it was given only supports "roughly forty thousand" or "up around a sixth." The extra digits are not knowledge; they are decoration that borrows the authority of precision. Treat a number that is more precise than its source could justify as a flag, not a finding — somewhere between the data and the answer, certainty got manufactured.

Getting the Numbers Right in Practice

Knowing that the arithmetic can be wrong is half the job; the other half is the handful of moves that catch it. These cost almost nothing and turn a risky habit into a safe one.

Ask it to show every step. "Don't just give me the answer — show the calculation, line by line, with the numbers you used." A bare total hides where it went wrong; a worked-out computation puts the error somewhere you can see it. When the percentage is computed against the wrong base or two figures get transposed, it shows up in a step, and you spot it by reading. This is the single most effective habit for quantitative work, and Think mode (Chapter 8) makes the steps fuller and more inspectable.

Push the actual computing onto a tool when you can. A language model reasoning about numbers in prose is fundamentally different from a calculator running them, and the difference is reliability. Where Grok can run or call out to a tool for the math, or where you can paste the figures into a spreadsheet and check the total yourself, do — let the model do the reading and structuring it is good at, and let arithmetic be done by something built for arithmetic. The honest framing is that Grok's bottom-line number is a claim until something deterministic has confirmed it.

Restate the question back before trusting the answer. A surprising share of analysis errors are not arithmetic at all — they are Grok answering a slightly different question than you asked, because the framing was ambiguous. Ask it to restate what it is computing and against what, and you catch the "growth versus loss," "this quarter versus year-over-year," "gross versus net" confusions before they become wrong conclusions. The misread in the opening story was exactly this kind: the number was transcribed perfectly and pointed the wrong way.

What You Can Actually Hand Grok

A practical note that shapes the whole chapter: analysis is only as good as what Grok can take in, and the intake has edges worth knowing.

Pasted text and documents are the strongest path — Grok's large context window (Chapter 4) lets it hold a great deal at once, so a long report, a contract, a transcript, or a stack of notes can go in whole and be reasoned over as a unit. Tables and figures in a document it can read, with the vision caveats from Chapter 15 riding along for anything it has to read off an image rather than parse as text. For genuinely large datasets — a spreadsheet of thousands of rows — you are usually better summarizing or sampling before you hand it over, both because of intake limits that drift over time (**check the current docs**) and because the model reasons better over a focused slice than a firehose. The durable habit: give Grok the relevant data, framed and bounded, rather than everything you have and a hope it sorts it out.

The Verification Habit

Because this matters so much at work, here is the actual discipline, stated as a habit rather than a warning. The thing that makes analysis with Grok safe is not avoiding it — it is genuinely fast and genuinely useful — but pairing it with a verification step proportioned to the stakes.

For anything that feeds a decision, verify the load-bearing numbers yourself. Not all of them, every time — that would defeat the purpose — but the ones the conclusion actually rests on. A figure that will sit in a board deck or drive a budget gets independently confirmed. Full stop.

Make the reasoning visible and read it. For real analysis, use Think mode (Chapter 8) and actually read the steps. The value is not that the reasoning is always right — it is that visible reasoning is checkable reasoning. You can spot the backwards interpretation or the wrong base in the steps far more easily than in a polished conclusion.

Trace conclusions to sources. Especially with live data (Chapter 10), glance at where the claim came from. Grok's sourced output makes this fast; the point is to actually do it for anything that matters.

Treat confidence as noise, not signal. Grok states the wrong answer with exactly the same assurance as the right one. Its tone tells you nothing about its accuracy. Calibrate your own trust to the verification you have done, never to how sure the output sounds.

None of this is heavy. For a quick read of a long document, a glance is enough. For a number that will be quoted in a meeting, you check it. The skill is matching the verification to the consequences — light where the cost of error is low, rigorous where a wrong figure becomes a wrong decision.

Used with that habit, Grok is a genuinely powerful analytical partner: it reads what you do not have time to read, structures what you have not had time to think through, and — uniquely — reasons over what is happening right now. It does not replace the analyst. It clears the runway so the analyst spends their attention on judgment instead of on extraction, which is exactly where a person's judgment is worth most.

Chapter 18: Coding with Grok

Paste a function and the error it keeps throwing into Grok, and it will often find the bug in seconds — the variable shadowed in an inner scope, the off-by-one, the thing a fresh pair of eyes catches and a pair that has read the same twelve lines for an hour does not. It explains why the code broke and hands back a corrected version. Then comes the move that is the whole chapter: you read the fix, understand it, and only then apply it. When the fix is right, the reading costs you a minute. On a different day, with a different bug, that same minute is what saves you from a fix that looked right and was not.

This chapter is about using Grok to write, debug, and explain code. Grok is a capable coding assistant — competent across generation, debugging, and explanation, and a reasonable peer to the other frontier models on this work (Chapter 3). It is not a magic wand, it is not a substitute for understanding your own code, and it is not the same thing as a dedicated coding agent. This chapter is about using it well and knowing where its lane ends.

What Grok Does Well With Code

Three things, and they map to most of what you actually want from an assistant while programming.

Generating code. Describe what you want — a function, a script, a class, a query, a config — and Grok produces a working version, often a good one, and usually faster than you would write it from scratch. It is strongest on the common stuff: well-trodden languages, standard libraries, the kind of code that has a million examples behind it. Boilerplate, scaffolding, the routine plumbing that has to exist but does not need your cleverness — this is where generation saves the most time, because it is exactly the work you would rather not type.

Debugging. Paste in code and an error, or describe the misbehavior, and Grok is genuinely useful at spotting the cause. The fresh-eyes effect is real: the model is not tired, has not been staring at the file for an hour, and has seen thousands of versions of your bug. For the off-by-one, the missed edge case, the shadowed variable, the wrong type, the misused API — the everyday bugs — it is fast and frequently right.

Explaining. This is underrated and may be the single most valuable coding use. Hand Grok an unfamiliar block — a function in a codebase you inherited, a regex you cannot parse, an idiom in a language you are learning, an error message that makes no sense — and ask it to explain. It walks you through what the code does and why, which is leverage for understanding rather than just producing. For working in unfamiliar territory, the assistant that helps you *understand* is often worth more than the one that writes for you.

Three Coding Tasks, Done Well

The categories are clearer with the work shown, so here is each in motion, with the technique that makes it actually pay off.

Generating: be specific or get generic. "Write a function to validate an email" gets you something — but probably not the something you needed, because the spec is in your head, not the prompt. The version that comes back ready to use looks more like: "Write a Python function that takes a string and returns True if it's a plausibly valid email, False otherwise. I want it to reject empty strings and obvious garbage but I don't need full RFC compliance — keep it readable, and add a docstring." Every choice you did not specify is a choice Grok makes for you, and its defaults are not your defaults. The discipline of describing the inputs, the outputs, the edge cases you care about, and the constraints you are under is the difference between a draft you adapt and a draft you discard. Treat the prompt like a small spec, because that is what it is.

Debugging: give it everything, not just the error. The fresh-eyes effect is strongest when Grok can see what you can. Pasting only the error line forces it to guess at the surrounding code; pasting the function, the error, and a sentence about what you *expected* to happen versus what *did* turns a guess into a diagnosis. "This function should return the user's most recent order, but it's returning the oldest one — here's the function and here's how I'm calling it." Now it can compare intent to behavior, which is exactly how a human debugger works. The more faithfully you reproduce what you can see, the more often the fix is real rather than plausible.

Explaining: ask the dumb question on purpose. The highest-value move in unfamiliar code is to stop pretending you follow it. "Explain this regex character by character." "Walk me through what this function does, line by line, like I've never seen this language." "Why would someone write it this way instead of the obvious way?" That last question is the one that turns explanation into learning — it surfaces the *reason*, not just the mechanics, and the reason is what lets you change the code safely later. Used this way, Grok is less a code generator than a patient senior engineer who will explain the same thing three different ways until it clicks.

A Fourth Use Worth Naming: Translating and Refactoring

Two more coding tasks deserve a line, because they sit between generation and explanation and are quietly some of the most reliable things Grok does.

Translating between languages and idioms. "Here's a function in Python — give me the equivalent in TypeScript." "Rewrite this loop as a list comprehension." "Convert this callback-style code to async/await." These work well precisely because the logic already exists and is correct; Grok is re-expressing known-good behavior in a new form rather than inventing behavior from a description. The error surface is smaller, because the hard part — getting the logic right — was already done by whoever wrote the original. The caveat is narrow but real: idiom translation can silently change semantics at the edges (a Python integer division becoming a JavaScript float, a truthy check that treats zero differently across languages), so the test still matters. But as a category, translation is one of the higher-confidence uses.

Writing the tests, not just the code. Ask Grok to "write tests for this function, including the edge cases you'd worry about," and it is often better at enumerating boundary conditions than at handling them in the original code. This is a genuinely useful inversion: the model's habit of pattern-matching against thousands of similar functions makes it good at remembering the empty list, the null, the off-by-one boundary, the duplicate input — the cases you forgot. You read the tests it proposes, keep the ones that matter, and now you have a net that catches exactly the subtle-edge-case failures the rest of this chapter warns about. Using Grok to generate the test that catches Grok's own bug is a quietly powerful loop.

The Limits, Stated Plainly

A guide that sold Grok as a flawless coder would be lying to you, and the honesty matters more here than almost anywhere, because broken code has a way of looking exactly like working code right up until it runs.

It can be confidently wrong. Grok will produce code that looks correct, reads correct, and is subtly broken — an edge case unhandled, a condition inverted, an API used in a way that compiles but misbehaves. The fluency is the trap. Plausible-looking code carries no guarantee of correct code, and the model's confidence tells you nothing about whether it works.

It can invent things. Asked to use a library, Grok may call a function that does not exist, pass an argument the real API never took, or assume behavior that is not the behavior. It is reconstructing a plausible API from training, not reading your installed version, and plausible is not the same as real.

It does not know your specifics. Grok does not see your full codebase, your exact dependency versions, your conventions, your constraints — only what you put in front of it. Code that is correct in the abstract can be wrong for *your* project: the wrong pattern for your architecture, a version mismatch, a violation of a convention it had no way to know. Context you did not provide is context it does not have. This is the failure mode that bites hardest for people working in a large existing codebase rather than a fresh script, because the abstractly-correct answer can quietly cut against an established pattern — reimplementing a helper that already exists three files over, reaching for a library your team deliberately banned, handling an error in a style nothing else in the project uses. The code passes review on its own terms and is wrong on the project's terms, and the only fix is to feed Grok the relevant context up front or to catch the mismatch yourself in review. The model cannot honor conventions it was never shown.

Currency cuts both ways. Grok's real-time strength means it can pull genuinely current information — a library's latest docs, a just-reported issue, a recent change — which is a real edge for fast-moving ecosystems (Part III). But the model's own coding knowledge still has the limits any model has, and "I found a recent post about it" is not the same as "this is correct for your setup." Live search helps you find current answers; it does not relieve you of vetting them.

Where the real-time edge genuinely earns its keep in coding is the breaking-change problem. Fast-moving libraries deprecate functions, rename arguments, and change defaults between versions, and a model frozen at a training cutoff will confidently hand you the old way of doing something long after it stopped working. Grok asking "what's the current way to do X in version Y" and pulling a recent answer is a real advantage here — the kind of thing that saves you from an hour of debugging code that was correct two releases ago. But the caveat sharpens rather than softens: a recent forum post or a fresh blog entry is *current*, not *correct*, and certainly not correct for your exact version and setup. Live search narrows the gap between the model's knowledge and today; it does not close the gap between today's internet and your specific machine. The official documentation for the version you actually have installed remains the authority. Use the live read to find candidates fast; confirm them against the real docs before you trust them.

The Read-the-Code Discipline

This is the non-negotiable, and it is the coding version of the beat that runs through every chapter in this part: **never run code you have not read and understood.** Not "skimmed." Read, and understood well enough to know why it works.

The temptation is obvious and strong. The code looks right, it is right there, running it is one click, and reading it carefully is work. But code you do not understand is a liability you have chosen to accept blind — and the failure modes above mean that liability is real, not theoretical. The cost of accepting broken code is not a polite error; it is a bug shipped into something that matters, a security hole, data corrupted, a problem that surfaces later when it is expensive.

So the discipline, concretely:

Read every line before you run it. Understand what it does and why. If you cannot follow a piece of it, that is precisely the piece to stop on — ask Grok to explain it, or work it out yourself, but do not run what you cannot read.

Verify the unfamiliar APIs. If Grok called a function or used a library you do not know cold, check the real documentation that it exists and behaves as claimed. This catches the invented-API failure directly.

Test it yourself. Run it, exercise the edge cases, confirm it actually does the thing. The model's confidence is not evidence; your test is. This is the entire safety net, and it is one you cannot delegate.

To make the stakes concrete, picture the classic subtle failure. You ask Grok for a function that calculates a discount, and it returns clean, readable code that handles the normal case perfectly. You read it, it looks right, you ship it. Three weeks later it turns out that when the discount is exactly 100% the function divides by zero, or when the input is negative it cheerfully returns a negative price, because the edge case never appeared in the examples Grok pattern-matched against. Nothing about the code *looked* wrong — it read as correct, and for the inputs you happened to test, it was. This is the shape of the danger: not code that fails loudly and immediately, but code that is right for the obvious cases and quietly wrong at the edges, shipped because "it looked right" stood in for "I checked it works." The test you write for the boundary case — the zero, the empty list, the negative number, the input twice as large as expected — is the thing that catches what reading alone cannot.

Iterate, Don't Accept

One more habit that separates productive coding-with-Grok from the frustrating kind: treat the first answer as a draft to push on, not a verdict to take or leave. If the generated function is close but uses a pattern you dislike, say so — "good, but use a list comprehension instead of the loop" or "this works but it's not handling the timeout case; add that." If a fix does not work, paste the new error back rather than starting over; the conversation carries the context, and the second attempt is usually better informed than the first. Coding with Grok is a dialogue, and the developers who get the most from it are the ones who steer through several rounds rather than judging a single shot. The corollary, of course, is that each round still gets read — iteration speeds you up, but it does not lower the bar on understanding what you finally run.

This discipline is also how you stay a programmer rather than a person who pastes. If the work is to land in something real, you are responsible for it — and you cannot be responsible for code you do not understand. The assistant that helps you understand more, faster, is a gift; the one that lets you ship what you cannot read is a trap wearing the same face.

Where Grok Fits Beside Dedicated Coding Tools

Worth being clear-eyed, because the landscape has more than one kind of tool, and the sibling books in this series cover the others in depth. There is a category of dedicated coding agents — tools that live in your editor or your terminal, read your whole repository, run your tests, and execute multi-step changes across many files as autonomous agents. Grok, in the chat interface this book is mostly about, is a different and lighter thing: a capable assistant you bring code *to*, in a conversation, rather than an agent you turn loose *in* your codebase.

That difference shapes what each is for, and the honest framing is that they are complementary, not competing. Grok-in-chat shines for the focused, conversational coding task — write this function, debug this error, explain this block, sketch this approach, talk through a design. It is fast, it is general, and it sits right beside the writing, analysis, and real-time work that fills the rest of your day, which is its own kind of convenience. A dedicated agent shines for deep, codebase-wide, multi-file autonomous work where the tool needs to *be in* your project to do its job. Many developers use both: Grok for the quick conversational question and the everyday explain-and-debug, a dedicated agent for the heavy in-repo work.

For the API side — building Grok's coding and reasoning capability *into* your own software rather than using it in chat — that is Part VI, and it is a different mode again.

The point of this chapter is the realistic middle: Grok is a strong, useful coding companion that makes you faster at writing, fixing, and understanding code, provided you stay the engineer in the loop and read what it gives you. Use it for the speed and the second pair of eyes. Keep the judgment, and the responsibility, with yourself — because the moment the code runs, it is yours.

Chapter 19: Everyday Productivity

Most real Grok use does not look like any of the headline features. It looks like this: a tangle of half-formed to-dos dumped into the chat with "help me sort out what actually matters this week"; a request, before a hard conversation, to play the other side so you can pressure-test your position; a five-line summary of a thread you do not have time to read; a quick "talk me through the tradeoffs" when you are stuck between two options. None of it is dramatic. All of it saves a few minutes and a little friction, and a few minutes and a little friction, several times a day, is what productivity actually is. This chapter is about that — Grok as the quiet everyday assistant rather than the showpiece.

The headline capabilities get the chapters and the demos. The everyday use gets the hours. It is worth taking seriously precisely because it is unglamorous: the steady low-stakes assistance that, summed across a working week, is where a tool like this earns or fails to earn its place in your day.

The Everyday Jobs

A handful of recurring uses cover most of it, and you will recognize all of them.

Planning and structure. Hand Grok a messy situation — a project with too many moving parts, a week with too much in it, a goal with no obvious first step — and ask it to help you structure it. Break this into phases. Sequence these tasks. What am I forgetting. It is good at imposing a sensible order on chaos, and a sensible order you can then adjust is worth far more than the blank intimidating mess you started with. The value is not that Grok's plan is perfect; it is that reacting to a draft plan is far easier than generating one from nothing.

Brainstorming. This is one of the genuinely best everyday uses, because it plays to exactly what the tool is and is forgiving of exactly where it is weak. Names, angles, approaches, gift ideas, ways to phrase something, things you might be missing — Grok is a tireless, fast, judgment-free generator of options. And here the it-might-be-wrong problem barely bites, because you are not trusting any single output; you are mining a pile of suggestions for the two or three worth keeping. Quantity is the point, and you supply the judgment. Grok's directness (Chapter 16) is a small bonus in this mode — it is more willing to throw out the blunt or unexpected option a more cautious tool would soften away.

Summarizing. The workhorse. Long email thread, dense article, meeting transcript, document you should read but will not — Grok condenses it. Reading and compressing is pure tedious-but-necessary work, the exact thing worth delegating, and the time it gives back is real and daily.

Thinking through decisions. Stuck between options, Grok is a useful thinking partner: lay out the tradeoffs, argue the other side, surface considerations you missed, stress-test your reasoning. It does not — and should not — make the decision; it widens your view before you make it. Asking it to argue *against* what you are leaning toward is a particularly good move, because it forces the case you were avoiding and a more willing-to-be-blunt model gives you a less padded one.

Drafting the small stuff. A surprising amount of a working day is tiny pieces of writing that are not hard but are friction: the calendar invite description, the one-line Slack update, the polite decline, the meeting agenda, the message you have been putting off because you cannot find the opening sentence. Grok clears these in a stroke. The value is not quality — most of these you could write yourself in two minutes — it is *activation*: it removes the small hesitation that turns a two-minute task into a thing you avoid for an hour. (Chapter 16 covers the writing craft; this is about the dozen micro-drafts that fill a day.)

Quick answers that would otherwise be a detour. "What's the difference between these two file formats." "How do I phrase this in French." "What's a polite way to ask for an extension." "Remind me how this thing works." None of these is worth a full research session, but each would otherwise be a tab-opening, search-and-read detour that breaks your focus. A fast capable answer in the chat you already have open keeps you in the task. The judgment caveat from later in the chapter applies — a quick answer can be quickly wrong — but for low-stakes orientation, the speed is the whole point.

Reformatting and converting. A surprising slice of friction is not thinking at all — it is the mechanical work of moving information from the shape it is in into the shape you need. Paste a wall of notes and ask for it as a clean bulleted list. Hand it a messy list of names and emails and ask for them as a comma-separated line. "Turn these meeting notes into a calendar-invite description." "Rewrite this as a table I can paste into a doc." None of this requires judgment, which is exactly why it is forgiving and exactly why it is a steady everyday win — there is little to get wrong in reformatting text you supplied, and the time it saves on the dozen small conversions a week is real. (Keep an eye on it the moment the conversion involves *interpreting* rather than reshaping — pulling dates out of prose, say — because that is where a reformatting task quietly becomes an extraction task with extraction's error rate.)

Remembering the thing you half-know. "What's that principle about the last 20% taking 80% of the time called." "There's a keyboard shortcut for splitting the window — what is it." "The word for when a brand name becomes the generic term." The tip-of-the-tongue retrieval that used to send you down a search rabbit hole resolves in one line. This is a small use that disproportionately preserves focus, because the alternative is not a thirty-second search — it is a thirty-second search that becomes a ten-minute detour through three other interesting things.

A Day With Grok in It

To make the everyday picture concrete rather than a list, here is what an ordinary morning can look like with the tool genuinely woven in — not a showcase, just the small uses stacking up.

You open the day with an inbox and a vague sense of dread, so you paste the subject lines and a two-line dump of what is on your plate and ask Grok to help you figure out the three things that actually matter today. It is not psychic — it does not know your politics or your real priorities — but reacting to its draft ordering is faster than building one from a blank page, and in two minutes you have a plan you have adjusted into something true. Mid-morning a long thread blows up and you do not have time to read forty messages, so you have Grok summarize it into "what happened, what's being asked of me, what's still open." Before a slightly tense call after lunch, you ask it to play the other person and push back on your position, and it surfaces the one objection you had been avoiding. Late in the day you are stuck on whether to take on a new project, so you have it argue the case against — the case you did not want to make to yourself — and you make the call clearer-eyed. None of these is a headline. All of them are five minutes and a little less friction, and the sum across the day is real. That accumulation, not any single feature, is what "everyday productivity" actually means.

Weaving In the Real-Time Advantage

This is what makes Grok's everyday productivity distinct from any other assistant's, and the natural fit is easy to miss until you have it. So much of daily work touches the present, and a frozen assistant simply cannot help with the present-flavored parts of an ordinary day.

When your planning depends on something current — what is happening in your industry this week, the state of a situation you are tracking, a development that just landed — Grok can fold live information into the everyday task (Part III). "Catch me up on what happened with X while I was out, then help me figure out what I need to do about it" is a single ordinary request that pairs real-time awareness with practical assistance, and only a real-time tool can take the first half of it. Brainstorming about a timely topic, summarizing not a document but *what is being said right now*, thinking through a decision shaped by current conditions — across the everyday jobs, the live edge means Grok is working with the actual present rather than a months-old snapshot of it. For a working life that runs on what is happening *now*, that is the difference between an assistant that is occasionally out of date and one that is current by default.

The simplest way to feel the difference is the "while I was out" problem. Everyone has come back from a few days away to a topic that moved without them — a project that shifted, a story in their field that broke, a debate that played out on X while they were offline. A frozen assistant is useless here by construction; it does not know the days you missed happened. Grok turns the catch-up into a question: "I've been out since Monday — what happened with [topic], and what are people saying about it now?" The answer is a fast, sourced read of exactly the window you missed, and you can chain straight into the practical half — "okay, given that, what should I prioritize this week?" — without ever leaving the chat. The catch-up and the planning, which used to be two separate chores, collapse into one short exchange. The same honesty from Part III rides along: the live picture is only as good as what Grok pulled, so anything load-bearing in your catch-up still wants a glance at the source before you act on it.

Making It a Habit Rather Than a Novelty

The gap between people who get real value from a tool like this and people who tried it twice and drifted away is almost never about the tool. It is about whether using it became a reflex. The everyday uses only pay off if you actually reach for them in the moment of friction, and that takes a little deliberate habit-building at the start.

Keep it within arm's reach. The single biggest predictor of whether you use Grok for the small stuff is whether opening it is frictionless — a tab already open, the app on your home screen, a keyboard shortcut. If reaching for it is a chore, you will do the two-minute task yourself and never form the habit. If it is already there, the reflex builds on its own. (Chapter 6 covers where Grok lives — web, mobile, on X — so you can put it wherever your friction is.)

Notice the moments of hesitation. The tasks worth handing to Grok announce themselves as small stalls: the email you keep not starting, the thread you keep not reading, the decision you keep circling. For the first couple of weeks, treat each of those stalls as a cue — "this is a Grok moment" — and ask. The point is not to use it constantly; it is to retrain the instinct so that the stall triggers the ask instead of the avoidance.

Let it learn your patterns. Over repeated use, the way you like things — your tone, your formats, the level of detail you want — becomes something you can establish rather than re-specify every time. Telling Grok how you prefer summaries (three bullets, no preamble) or how you like decisions framed (tradeoffs first, recommendation last) once, and reusing that, turns each interaction from a fresh negotiation into a quick handoff. The everyday value compounds when the tool is shaped to your defaults.

Start with the forgiving uses. If you are building the habit, lead with brainstorming, drafting, and summarizing — the refine-and-go tasks where a wrong output costs you nothing and the speed is pure gain. Those build trust and reflex without exposing you to the verification burden that the trust-and-verify tasks carry. Once reaching for Grok is automatic on the safe stuff, you will have the judgment to know when *not* to on the rest, which is the subject of the next section.

Batch the small stuff instead of scattering it. A subtle habit that experienced users converge on: rather than interrupting deep work to fire off a tiny request the moment it occurs to you, let the small tasks pool and clear them in one pass — the three micro-drafts, the two summaries, the reformatting job — at a natural break. The reason is not the tool; it is your attention. Each context-switch into and out of focused work costs more than the task it served, so the assistant that is supposed to reduce friction can paradoxically *add* it if every small use yanks you out of what you were doing. The reflex you are building is "this is a Grok moment," but the discipline that makes it pay is "and I'll do it at the next seam, not right now." Frictionless access and deliberate batching are not in tension — together they are how the small uses add up without fragmenting the day they were meant to smooth.

A Realistic Picture of Where It Helps — and Where It Doesn't

Here is the honest part, because a productivity chapter that promised Grok saves time on everything would be selling, not guiding.

Where it genuinely saves time: the blank-page problem, in all its forms — the first draft, the initial plan, the starting list of ideas. Tedious compression — summarizing, condensing, extracting. Getting unstuck — a thinking partner when you are spinning. Quick capable answers to questions that would otherwise be a search-and-read detour. And, distinctively, anything where being current is the point. These are real, repeatable savings, and they are most of why the tool is worth having.

Where it does not, and pretending otherwise wastes your time: Anything where verifying Grok's output costs more than doing the work yourself. If a task is quick for you and checking the assistant's version would take longer than just doing it, reaching for Grok is overhead, not help — the assistant is a net loss on work that is faster done than verified. Anything requiring judgment, taste, or context only you have — Grok can inform those, but handing them over wholesale produces generic output you then have to fix, which is often slower than starting yourself. And anything where a confident wrong answer is worse than no answer: a high-stakes fact, a number that feeds a decision, a claim you will repeat as your own. There, the verification the output demands (Chapters 16 and 17) can swamp the time it saved.

The skill is sorting one kind of task from the other, and it is the same sort that runs through this whole part. **Where Grok's output is a starting point you will refine, it is leverage. Where its output is a finished answer you must trust, it is a liability you have to verify — and sometimes**

verifying costs more than doing. Brainstorming, drafting, summarizing, structuring: refine-and-go, almost always a win. A figure for the board deck: trust-and-verify, and the verification is the real work. Learn to feel which one you are in, and you will reach for Grok exactly when it helps and leave it alone when it does not.

That judgment is the quiet through-line of the entire "for work" part. Grok is a genuinely capable everyday assistant, made distinct by a real-time awareness none of its peers can match, and most useful to the person who knows precisely what to hand it and what to keep. It saves the minutes that add up. It does not replace the thinking — and the parts of your day that are actually thinking are the parts worth protecting from it.

That care about *where* the tool helps and where it quietly hurts is the right frame to carry into Part VII, which turns from what Grok does well to what it gets wrong — the failures, the biases, the privacy questions, and what responsible use at work really asks of you.

Part VI: Building on Grok

Audience: developers and builders, and the technically curious. The deep parts are signposted — navigate around them if code isn't your work.

"The chat box is one way to reach Grok. The API is how you put its sense of the present inside something you built."

— A working principle

Chapter 20: The Grok API

A friend who runs a small product team called me last spring with a problem that turned out to be the whole reason this Part exists. His app helped people track a fast-moving regulatory area, and his users kept asking it questions about things that had happened that week — and the AI inside his product kept shrugging, because it was a frozen model that knew nothing past its training cutoff. "It's smart," he said, "but it's stuck in the past, and the past is the one thing my users don't need." What he wanted was not a smarter model. It was a model that could *look*, on demand, at what was true right now, from inside his own software. That is exactly what the Grok API is for, and this chapter is about getting oriented to it.

Everything in this Part is written for people who build — developers wiring Grok into their own applications. If you only ever use Grok through the chat product, you can skip Part VI cleanly and lose nothing; the rest of the book is yours. But it is worth knowing this surface exists, because it is where Grok stops being a tool you open and becomes a capability you build with. And one caution belongs at the very top, the same one Chapter 4 raised about model numbers: API details — endpoints, parameter names, exact pricing, limits — are the most perishable material in the book. I will describe the durable shape and flag every specific. **The authority is always xAI's current API documentation, not this page.**

What the API Actually Is

The Grok API is a programmable doorway to the same models that power the chat product. Instead of a person typing into a box on grok.com (Chapter 6) and reading the answer, your program sends Grok a request over the network — a prompt, some context, some settings — and receives the model's response back as data your code can use. The intelligence is identical; what changes is who is doing the asking. In the chat product, a human is in the loop. Through the API, your software is.

That distinction is the entire point. It means Grok's capabilities — the reasoning, the writing, the analysis, and most importantly its real-time search — can live *inside* applications you write: a customer-support bot that knows about today's outage, a monitoring dashboard that summarizes live sentiment, an internal tool that drafts on demand. The chat product is Grok for people. The API is Grok for programs.

The shift in who asks changes more than it first appears. A human using the chat product reads every answer, notices when something looks off, rephrases when the first reply misses, and quietly does a hundred small acts of judgment that never show up in the transcript. Your software does none of that unless you build it in. When you move a workflow from the chat box to the API, you are not just changing the interface — you are removing the person who was silently catching problems, and taking on the job of catching them yourself in code. That is the deal the API offers: more power, more reach, and more responsibility, all at once.

Who It's For

Three rough groups reach for the API, in my experience.

Builders embedding Grok in a product — a startup or a team putting conversational AI into an app, where Grok is the engine behind a feature their users touch. This is the largest group and the one this Part is mainly written for.

Automators and scripters — people wiring Grok into pipelines, batch jobs, and internal tooling, where there is no user-facing chat at all, just Grok doing a step in some larger process on a schedule or in response to an event.

Tinkerers and evaluators — developers kicking the tires, comparing Grok against other providers, prototyping an idea to see whether the real-time edge changes what they can build. Many serious deployments started here.

If none of those is you, this is a fine place to rejoin the book at Part VII. If one of them is, read on.

Getting Oriented: Keys, Auth, and Endpoints

The mechanics of using any modern AI API follow a shape you may already know from other providers, and Grok is no exception. I will keep this conceptual on purpose, because the precise strings are exactly what drifts.

You authenticate with an API key. Programmatic access uses a key — a secret credential you generate from xAI's developer console and attach to each request, rather than the interactive sign-in a human uses on the website. The key identifies your account, meters your usage, and is the thing you must guard. Treat it like any other secret: store it in environment variables or a secrets manager, never paste it into client-side code or commit it to a repository, and rotate it if it leaks. A leaked key is someone else spending your money, and worse. **Generate and manage keys from xAI's current developer console; confirm the exact setup steps there.**

The "never put it in client-side code" rule deserves a beat of its own, because it is the mistake new builders make most. It is tempting, when you are wiring Grok into a web app or a mobile app, to call the API straight from the browser or the phone with the key embedded in the front-end code. Do not. Anything that ships to a user's device is readable by that user, and a key sitting in JavaScript that loads in a browser is a key you have effectively published to the world — people scrape exactly these. The durable pattern is to keep the key on a server you control: the user's device talks to your backend, your backend holds the secret and talks to Grok, and the key never leaves your infrastructure. That indirection also gives you the place to put your own rate limiting, logging, and spend caps, which you will want anyway.

You send requests to an endpoint. Your code makes network calls to xAI's API — a base URL with paths for the operations the API supports, the central one being a chat-style request where you send a list of messages and get the model's reply. The request carries your key, the model you want, your messages, and settings like how long the answer can be. **The exact base URL, paths, and parameter names live in xAI's docs and are check-the-docs details — do not hardcode them from memory or from this book.**

The shape will feel familiar. Grok's API is broadly industry-standard in its design — a chat-completions style of interaction that developers coming from other AI providers tend to find recognizable, sometimes to the point where existing client libraries can be pointed at it with minimal changes. I want to be careful here, because "compatible" is a word that gets oversold: the degree of drop-in compatibility, which libraries work, and which features map cleanly are precisely the things that change release to release. So hold the durable claim — *the API follows familiar, conventional patterns* — and verify the specific compatibility you are counting on against the current documentation rather than trusting a blanket promise.

In practice, that familiarity has a real payoff and a real trap. The payoff is that if your code already speaks to another provider through a standard client library, getting a first Grok response back is often a matter of changing a base URL, swapping in your xAI key, and naming a Grok model — sometimes minutes of work rather than days. The trap is assuming that because the *base* call ports cleanly, everything ports cleanly. The features that make Grok distinctive — live search above all (Chapter 21) — are exactly the ones least likely to have an identical shape in another provider's library, because no other provider offers them in the same way. So expect the plain chat request to feel like home, and expect the interesting parts to need their own reading of xAI's docs.

A Minimal Illustrative Request

To make this concrete, here is the *shape* of a basic request in pseudocode. I am writing it loosely on purpose — this is meant to show you the moving parts, not to copy-paste. The real parameter names, the real endpoint, and the real client library belong to xAI's current docs, and Appendix B walks through a working quickstart you can actually run.

```
# ILLUSTRATIVE ONLY – verify every name against current xAI API docs
POST  <xAI chat endpoint>
Header:  Authorization: Bearer <YOUR_API_KEY>
Body:
  model:     "<current flagship model name>"   # e.g. a Grok 4.x model
  messages:  [
    { role: "system", content: "You are a helpful assistant." },
    { role: "user",   content: "Summarize this report: ..." }
  ]
  max_tokens: <a cap on the reply length>

# Response comes back as structured data containing the model's reply.
```

Four things to notice, because they are durable even as the names drift. You pick a **model**. You send a list of **messages** with roles — typically a system message that sets behavior and a user message that asks. You set **limits** like a maximum reply length. And you get back **structured data**, not loose text, so your program can act on it. Every modern AI API is some version of this, which is the good news: if you have used one, Grok will not feel foreign.

The **system message** is worth lingering on, because it is the lever new builders most often underuse. In the chat product, a person re-steers the model conversationally — "no, shorter," "more formal," "just give me the number." Through the API there is no person to re-steer; the system message is where you set, once and for every request, who the model is and how it should behave: its tone, its format, the boundaries of what it should and should not do, the persona your product wants it to wear. A vague system message ("You are a helpful assistant") gets you a generically capable model. A specific one ("You are the support assistant for a travel app. Answer in two sentences. If you are unsure, say so and offer to escalate. Never invent booking details.") gets you a model that behaves like part of your product. It is the cheapest, highest-leverage piece of the request, and it is the one you will iterate on most.

The **structured data** in the response matters for the same reason. Because you get back data rather than a wall of text, your code can read not just the model's words but the surrounding facts — whether the model finished its answer or got cut off by your length cap, how many tokens the call consumed (which is your cost, Chapter 22), and, when you use tools, *what* the model is asking you to do rather than a reply at all (Chapter 21). A program that reads only the text and ignores the rest of the response is throwing away the information that lets it behave well. Read the whole envelope, not just the letter inside.

Choosing a Model

The API exposes the Grok model family (Chapter 4), and choosing among its members is the first real decision you make. The trade-offs are the ones that chapter laid out, now seen from a developer's seat.

The flagship — Grok 4.3 as of this writing — for hard work. When the task needs the strongest reasoning, the deepest analysis, the most reliable output, you reach for the top model and pay the top rate. **The current flagship name is a check-the-docs detail; confirm it before you wire it in.**

Lighter or faster members for volume and routine work. Not every API call needs the flagship. High-volume, latency-sensitive, or simple tasks are often better served by a faster, cheaper member of the family, and at scale that choice is the difference between a sustainable bill and a painful one. Chapter 22 is the honest accounting of cost; the short version is that model choice is your biggest cost lever.

The way to make that choice well is to resist deciding it by gut. The instinct — "I'll just use the best model so the answers are good" — feels safe and is usually wrong at scale, because the best model is the most expensive and the slowest, and a great many tasks do not need it. Consider a product that does three things: classifies an incoming support message into a category, drafts a polite reply, and, on hard tickets, reasons through a thorny multi-part problem. Those are three different difficulties wearing one product's skin. Routing the classification to a light, fast member, the routine drafting to a mid member, and only the genuinely hard reasoning to the flagship can cut cost dramatically while *improving* the experience, because the light model also answers faster. The discipline is to match the model to the difficulty of each task, not to the importance of the product — and the only honest way to know which model is good enough for a given task is to try the cheaper one on real examples and look at whether it actually fails.

The durable advice mirrors the chat product's: **default to a capable model, drop to a lighter one where the task allows, and reach for the flagship where the difficulty actually lives.** The exact roster of named members shifts with every release, so treat the *idea* of a family as stable and the *specific members* as something to look up.

How the API Relates to the Chat Product

It helps to be clear about what the API is and is not relative to grok.com and the apps (Chapter 6). Same models, different doorway — but a few things genuinely differ, and they matter.

The API is **metered separately** from a consumer Grok subscription. Paying for a Grok subscription as a person does not generally cover programmatic API usage; the API bills on its own pay-per-use meter, by the token. Know which meter you are on before you build something that runs often (Chapter 22). Consumer pricing is Chapter 5's subject and stays there; API token pricing is this Part's, and I will give you those figures with verify-flags.

The API also gives you **fewer guardrails by default and more responsibility**. The chat product wraps the model in an interface, a history, and a set of safety and rate behaviors built for humans. The API hands you the raw capability and expects you to build the surrounding judgment — the input validation, the output checking, the cost controls, the user-facing guardrails. That is power and obligation in the same package, and Part VII's responsible-use material applies with extra force the moment a model is acting inside software rather than in front of a person reading every reply.

There is also a quieter difference worth naming: **the API does not remember.** The chat product keeps your conversation, so the model sees what you said earlier in the thread without you re-sending it. The API is, by default, stateless — each request stands alone, and the model knows only what you put in *that* request's messages. If you want the model to remember the conversation so far, you carry that history yourself, sending the relevant past messages back with each new turn. This is not a flaw; it is what gives you control over exactly what context the model sees. But it surprises builders who expect the API to "remember" the way the chat product does, and it has a cost dimension too — every past message you re-send is input tokens you pay for again (Chapter 22), so a long conversation gets more expensive with every turn unless you manage what you carry forward.

What the API does *not* change is the model's character: its strengths, its failure modes, and above all its real-time edge are the same through the API as in the chat box. Which brings us to the part that makes building on Grok genuinely distinctive rather than just another API integration — exposing Grok's live web and X search to your own application, so the software you ship can answer questions about the present, not only the past. That capability, and the function-calling machinery underneath it, is where we go next. And for the hands-on, run-it-yourself version of everything sketched here — keys, a real request, a first response — Appendix B is the quickstart to keep open beside the docs.

Chapter 21: Tool Use and Live Search Tools

Wire a model into an application that can *act* rather than just talk, and the first thing that surprises you is small and slightly unnerving: you ask it a question, and instead of answering, it stops and says — in effect — "I need to look something up; please run this lookup for me and hand me the result." It has not answered. It has asked *your code* to go fetch something, intending to use what comes back. That hand-off is the heart of tool use, and once you see it, you understand how a language model stops being a clever text generator and becomes the reasoning core of a real application. This chapter is about that mechanism, and then about the version of it that makes Grok stand apart: turning its live web and X search into a tool your software can call.

This is builder territory, like the rest of Part VI. If you do not write code, the durable idea is still worth carrying — *Grok can be wired to reach into the live present from inside other people's apps* — and you can skip the mechanics with a clear conscience.

What Tool Use Is

A language model, on its own, can only produce text. It cannot check a database, call your API, look at a live price, or read the web — it can only *say things*. Tool use, also called function calling, is the bridge across that gap. You describe to the model a set of tools it is allowed to use — each one a function in your code, with a name, a description of what it does, and the inputs it expects. Then, when the model decides a tool would help answer the question, instead of guessing it emits a structured request: "call this tool, with these arguments."

Here is the part that surprises people: **the model does not run the tool.** It cannot. It only asks. Your code receives the request, runs the actual function — querying your database, hitting your API, whatever the tool does — and passes the result back to the model, which then folds that real information into its answer. The loop looks like this, conceptually:

```
# CONCEPTUAL FLOW — names and shapes vary; verify against current xAI docs
1. You send:     the user's question + a list of tools the model may use
2. Model replies: "call tool `lookup_order` with { id: 4471 }"   (not an answer)
3. Your code runs lookup_order(4471)  →  gets the real data
4. You send back: the tool's result
5. Model replies: a final answer that uses the real data
```

That round-trip — model asks, your code acts, model uses the result — is the whole pattern, and it is what lets a model that can only produce text nonetheless operate on the real world. You define the tools; the model decides when to reach for them; your code stays in control of what actually runs. **The exact way you declare tools and read back the model's requests is a check-the-docs detail and follows broadly conventional patterns; confirm the current shape before you build.**

The Tool Description Is the Whole Game

There is a non-obvious skill buried in that loop, and it is worth pulling out because new builders trip on it constantly. When you give the model a tool, the thing it reads to decide *whether and how* to use that tool is the tool's **description** — the plain-language sentence you write about what the tool does. The model does not see your code. It sees your description. So the description is not documentation for other programmers; it is a prompt aimed at the model, and its wording directly shapes behavior.

A vague description gets you unreliable tool use. Tell the model a tool does "order stuff" and it will call that tool at the wrong times, or fail to call it when it should, or pass the wrong arguments — and you will spend an afternoon confused about why. Tell it the tool "looks up the status, ship date, and contents of a customer order given the numeric order ID; use this whenever the user asks about an existing order" and the behavior sharpens immediately, because you have told the model exactly when this tool is the right move and exactly what it needs to provide. The same care applies to the inputs: name and describe each argument so the model knows that `id` is a numeric order number and not, say, a tracking code. Treat tool descriptions as something you write deliberately and revise after watching how the model actually calls them. It is the single highest-leverage piece of wiring a tool, and it is pure prose, not code.

A Loop, Not a Single Call

The five-step flow above hides something that matters for both correctness and cost: it can repeat. A model working through a real task may call one tool, read the result, decide it now needs a *second* tool, call that, read again, and only then answer. A question like "did my last order ship, and if so when will it arrive at my address?" might have the model call an order-lookup tool, then a shipping-estimate tool with the address it got back, then finally answer. Each of those round-trips is another request to Grok and another helping of tokens (Chapter 22). This is what people mean by an *agentic* application — the model is not answering once, it is running a little loop of think-act-observe until the job is done.

That loop is powerful and it is where the cost and the failure modes both live. Powerful, because it lets the model accomplish multi-step tasks you could never script by hand, adapting its path to what each tool returns. Risky, because a loop that does not terminate, or that calls an expensive tool more times than you expected, runs up cost and latency fast — and because every tool result you feed back is something the model will trust, so a tool that returns bad data leads the model confidently astray. The durable engineering habits: cap how many tool-call rounds a single request may run, so a confused model cannot loop forever; validate what your tools return before handing it back; and watch the token consumption of multi-step sessions, because they are the expensive end of the spectrum.

The reason all of this matters for Grok specifically is that xAI ships a tool you would otherwise have to build yourself — and it happens to be the one tool that plays directly to Grok's defining strength.

Grok's Standout: Live Search as a Tool

Everything Part I and Part III said about Grok comes down to one capability: it can look at what is true *right now*, drawing on real-time web search and the live firehose of X (Chapter 10). In the chat product that shows up as DeepSearch and live answers (Chapter 9). The thing that makes the Grok API genuinely distinctive — not just another competent chat API — is that it exposes that same live search as a tool your application can call.

Sit with what that unlocks. Most AI you can build into a product is frozen at a training cutoff; ask it about this morning and it apologizes. Wire Grok's live search into your app, and your software can answer questions about the present — today's news, this hour's sentiment, the reaction to a thing that shipped an hour ago, whether an event has actually happened yet. The signature real-time edge that makes Grok worth choosing as a *person* becomes a building block you assemble into software for *other* people. That is the API turning Grok's whole reason for existing into infrastructure.

Concretely, the live-search capability lets you have Grok consult current web and X sources as part of answering an API request — either as something you switch on for a request, or as a tool the model invokes when it judges the question needs fresh information. **The exact mechanism — what it is called, how you enable it, what controls you have over which sources it draws on — is precisely the kind of detail that changes between releases, so treat the capability as durable and verify the specifics against xAI's current API documentation.** The durable claim is the one that holds: through the Grok API, you can give your application a live read on the present, and almost nothing else in the market offers that in the same first-class way.

Two Ways to Reach for the Present

Worth distinguishing the two postures you can take toward live search, because they trade off cost against responsiveness, and the right choice depends on your product.

You decide, per request, whether to search. Your application knows things about the question before the model sees it. If a user is asking "summarize this document I pasted," there is nothing to search for, and switching live search off saves the cost and the latency. If a user is asking "what's the news on X today," your code can switch it on. This posture keeps you in control of when the meter runs, which is the cheaper and more predictable path — and for many products, your code can route well enough by simple signals about what kind of question came in.

You let the model decide. Offer live search as one of the tools the model may call, and let it judge — question by question — whether it needs fresh information. This is more flexible and handles the ambiguous middle better than a crude routing rule, because the model often understands the question's needs better than a keyword check would. It is also less predictable in cost, because now the model, not your code, is deciding when to spend on a search, and a model that reaches for search reflexively can run up surprises. Many real applications blend the two: your code makes the obvious calls and the model handles the judgment calls.

Neither is wrong. The honest guidance is that the more control you keep over *when* search fires, the more predictable your bill and the less your exposure to noisy live data on questions that never needed it — so default to deliberate, and hand the decision to the model where the flexibility genuinely earns its cost.

A Worked Conceptual Example

Picture a customer-facing assistant inside a travel app. A user types: "Is there anything happening at my destination this weekend I should know about?" A frozen model cannot answer — it does not know what *this weekend* holds. Here is how the Grok-with-live-search version goes, kept conceptual so the durable shape shows through:

```
# ILLUSTRATIVE — verify tool/search mechanics against current xAI docs
User asks:  "Anything happening in Lisbon this weekend I should know about?"

Your request to Grok:
  - the user's question
  - live search enabled (or offered as a tool the model may call)

Grok, internally:
  - recognizes this needs current information
  - searches live web / X for recent, relevant items
  - synthesizes: a festival, a transit strike, a weather alert

Grok returns:
  a current, sourced answer your app can show — ideally with where
  each claim came from, so the user (and you) can check it.
```

The application did not hardcode anything about Lisbon or this weekend. It handed Grok a live capability and let the model use it. That is the difference between software that knows only the past and software that can speak to the present — and it is buildable today through this one API.

Now run the same example forward into a less comfortable scenario, because the travel app makes a fine illustration of both halves of the trade. Suppose the live search surfaces an X post claiming a transit strike that, in fact, was called off an hour ago — or never confirmed at all. Grok faithfully reports what it found, your app faithfully shows it, and your user reschedules a trip around a strike that is not happening. Nothing in the pipeline malfunctioned. The model did its job, the search did its job, your code did its job — and the output was still wrong, because the *source* was wrong and nobody in the chain was positioned to know. That is not a hypothetical edge case; it is the ordinary texture of live data, and it is why the caveats below are not boilerplate.

The Honest Caveats

Two of them, and neither is optional.

Live search costs more, and it costs unpredictably. Every real-time search or tool call your application makes is work the API meters on top of the base token cost — and an agentic, search-heavy app burns through spend faster than one that just answers from the model's own knowledge. A single user question that triggers several searches is several times the cost of one that does not. Chapter 22 is the full accounting, but build with it in mind from the first line of code: the live edge is a metered privilege, not a free one, and an app that searches on every message can surprise you on the bill. Decide deliberately *when* your application reaches for live search rather than letting it fire on every request.

Live information is noisy, and the noise comes through the API too. This is the hazard Chapter 10 and Chapter 23 dwell on, and it does not soften because a program is in the loop instead of a person — if anything it sharpens, because there is no human reading every answer and raising an eyebrow at the dubious ones. The live web and X carry rumor, error, manipulation, and outright propaganda right alongside the truth, and a real-time tool will surface whatever is loud, not whatever is correct. When you ship live search inside your product, you are shipping that exposure to your users. So build the guardrails the chat product's human would otherwise provide: prefer answers that carry their sources so claims can be checked, be wary of using raw live output for anything consequential without verification, and tell your users when an answer rests on unvetted live information. The real-time capability cuts both ways, exactly as Chapter 1 warned — it is the most powerful thing your app can offer and the easiest way for it to mislead, and baking it into software that other people trust doubles the stakes of both.

There is a design principle that follows from both caveats and that I would put on the wall of any team building this: **surface the sources, and surface the uncertainty.** A live answer that arrives with its receipts — here is the post, here is the page this came from — lets your user (and your support team, and you) do the verification the model cannot. A live answer that arrives as a bare confident sentence has stripped out exactly the information needed to catch it when it is wrong. Designing your product to carry sources through to the user is not a nicety; on a real-time app it is the primary defense, and it is cheap to build in from the start and painful to retrofit after the first bad answer goes out.

Tool use is what turns Grok from a model into the brain of an application, and live search is the tool that makes building on Grok a different proposition from building on anyone else. Use it for what it is — a live line to the present, metered and noisy, that almost nothing else can give you — and design around both its cost and its trust the way the next chapter, and Part VII, insist you must. Cost is where we turn now, because the moment your software can call out to the live world on its own, spending stops being something you do and becomes something your code does for you.

Chapter 22: Cost and Rate Limits

The bill that taught me to respect API economics was not large, but it was startling in shape. I had built a small prototype that called a model for every item in a list, and a list I expected to be a few hundred items long turned out, because of a bug, to be a few hundred thousand. The model did exactly what I asked, several hundred thousand times, and the meter ran the whole way. Nothing was broken except my assumption that "it's just a few cents per call" stays small. It does not stay small when your code, rather than your fingers, is doing the asking. This chapter is about spending on the Grok API deliberately — what it costs, what makes it cost more, where the ceilings are, and how to stay in control.

I am going to do this the way Codex's pricing appendix does it, and the way Chapter 4 promised: I will give you the durable structure, and I will flag every number. **The structure here is reasonably stable. The figures are not, and the only authority on today's figures is xAI's current API pricing documentation — check it before you budget anything against it.** Treat what follows as the shape of the thing and the questions to ask, not as a quote.

Token Pricing: The Base Meter

API usage is priced by the **token** — the small chunks of text that models read and write, roughly a few characters each, with a token landing somewhere under a typical English word on average. Every request bills on two counts: the **input tokens** you send (your prompt, your system message, the conversation so far, any documents) and the **output tokens** the model generates back. You pay for both, at different rates.

For the flagship — Grok 4.3 as of this writing — the token pricing is in the neighborhood of **$1.25 per million input tokens and $2.50 per million output tokens**. Two things to read off those numbers. First, output costs more than input — about double here — which is standard across the industry and reflects that generating text is more expensive work than reading it. Second, the unit is *per million tokens*, which is why individual requests feel nearly free and why volume is where cost actually lives: a single question is fractions of a cent, but a million questions is real money, and software asks a lot of questions.

Verify both figures against xAI's current API pricing before you rely on them. API rates move, they differ by model within the family, and the flagship's rate is not the lighter members' rate. This pair of numbers is a reference point of the right order of magnitude, not a price I can stand behind by the time you read it — that is the deal this whole book makes about perishable figures, and pricing is the most perishable figure of all.

It helps to make the arithmetic concrete, because "per million tokens" stays abstract until you put a real workload through it. Imagine a support assistant that handles ten thousand conversations a day. Say a typical conversation feeds the model around two thousand input tokens — the system message, a little history, the user's question — and the model writes back around five hundred output tokens. At the illustrative rates above, that is roughly a quarter of a cent of input and an eighth of a cent of output per conversation: call it under half a cent each. Ten thousand of those a day is somewhere around thirty-some dollars a day, very roughly a thousand dollars a month, purely on base tokens. Notice what that exercise reveals: no single number in it is alarming, the per-conversation cost is trivial, and yet the monthly total is a real line item. That is the whole character of API cost — it is death by a thousand cheap calls, and the only way to see it coming is to multiply the cheap call by the volume *before* you ship, not after. **The rates are illustrative; the method is the durable part — estimate tokens per call, multiply by calls per day, and look at the monthly number with clear eyes.**

The single biggest lever on this meter is **model choice** (Chapter 20). Routing routine, high-volume work to a lighter, cheaper member of the family and reserving the flagship for the hard parts is, at scale, the difference between a bill you can live with and one you cannot.

Input Tokens Add Up Quietly

One part of the base meter catches people off guard, so it earns its own mention: you pay for input every single time, and input is easy to let balloon. Because the API is stateless (Chapter 20), an application that wants the model to remember a conversation re-sends the prior messages with every new turn — which means a long conversation pays for its growing history again on each message. Turn ten of a chat is not just turn ten's tokens; it is turns one through nine riding along as input, billed afresh. The same quiet inflation happens when you stuff large documents or generous context into every request "just in case." None of it is wrong, but it is cost, and it is the cost people forget to count because the visible thing they asked for was small. Trimming the history you carry forward, and sending only the context a given request actually needs, is a real and often overlooked lever — one that also tends to *improve* answers, since a model handed only what is relevant reasons more cleanly than one buried in everything.

What Search and Tool Calls Add on Top

Here is where Grok's economics get their own character, and where the prototype-bug lesson bites hardest. Token pricing is the base. But the capability that makes Grok worth building on — live web and X search (Chapter 21) — is additional work the API meters *on top of* the base token cost. A request that triggers real-time search is not just paying for the tokens in and out; it is paying for the search the model performed to answer.

The practical consequence is that an **agentic, search-heavy application costs meaningfully more than a plain question-and-answer one** — and it costs *less predictably*, because the model decides when to search, and a single user message might cause several searches or none. The same is true of tool use generally (Chapter 21): a model that calls tools and reasons across several turns to finish a task consumes far more than a model answering once from its own knowledge. Long, multi-step, tool-using sessions are the expensive end of the spectrum; single, self-contained answers are the cheap end.

The unpredictability is the part worth bracing for, because it is unlike the base meter in a way that breaks naive budgeting. Base token cost scales smoothly with traffic — twice the users, roughly twice the bill — so you can forecast it. Search-and-tool cost scales with *behavior*, and behavior varies. A run of ambiguous questions that each trigger three searches can cost several times what the same number of plain questions would, and you do not control that ratio directly when you have handed the search decision to the model. The defenses are the ones Chapter 21 raised, now seen through the cost lens: keep deliberate control over when search fires rather than letting it be reflexive, cap how many tool-call rounds a single request may run so a confused model cannot loop up a bill, and watch the agentic sessions specifically rather than trusting an average that the occasional runaway session quietly poisons.

The exact additional cost of live search and other tools is a check-the-docs figure, and the way it is charged can change — confirm it on xAI's current pricing page. The durable point needs no flag: the things that make Grok special through the API are the things that add cost, so the more you lean on the real-time edge, the faster the meter runs. Build knowing that, and decide deliberately *when* your software reaches for search rather than letting it fire reflexively on every message.

Rate Limits: The Other Ceiling

Cost is one constraint; **rate limits** are the other, and they bite differently. A rate limit caps how much you can call the API in a window of time — how many requests per minute, how many tokens per minute, that sort of thing. Exceed it and the API turns you away temporarily, returning an error rather than an answer until the window resets. This is not about your total spend; it is about your *pace*, and it is the constraint that surprises people who budgeted for cost but not for throughput.

Limits typically scale with your account standing — newer or lower tiers get tighter ceilings, established or higher-spending accounts get more headroom — and the specific numbers, the windows, and how to request an increase all **live in xAI's current documentation and console and are check-the-docs details.** What is durable is that the limits exist and that your software must expect them: a burst of traffic, a batch job, or a sudden spike of users can hit a rate ceiling, and code that does not handle that gracefully will fail in front of someone.

The failure is worth picturing concretely, because rate limits tend to be invisible right up until the worst possible moment. They do not bite during quiet development, when you are sending one request at a time and never come near a ceiling. They bite the afternoon your product gets popular — a launch, a mention that sends a crowd, a batch job that fires ten thousand requests in a tight loop — exactly when you least want it to fail. An application written on the assumption that the API always answers immediately will, on that afternoon, start throwing errors at real users while everything else looks fine. That is the trap: the code that ignores rate limits passes every test you run and fails only under the load you were hoping for.

The durable engineering response is to **build for limits from the start**: catch the rate-limit error specifically (it is a distinct, recognizable response, not a generic failure), back off and retry after a pause rather than hammering — and lengthen the pause on repeated failures rather than retrying at the same furious pace — and where you can, smooth bursts into a steadier stream instead of firing everything at once. For batch work, that means metering your own sending pace deliberately rather than dumping the whole job on the API in a heartbeat. An application that assumes the API will always answer immediately is an application that will break the first busy afternoon.

Managing Spend Deliberately

Pulling it together, a handful of durable practices keep API cost from becoming a surprise — the discipline matters more than any specific number.

Choose the cheapest model that does the job. This is the largest lever by a wide margin. Default to a lighter member for routine and high-volume work; reserve the flagship for tasks that genuinely need it.

Treat live search as a deliberate choice, not a default. Since search and tools add cost on top, decide in code when a request actually needs the live present, rather than letting every message trigger a search it did not require.

Cap what you can cap. Limit reply lengths where long output is not needed, trim the context you send rather than passing everything every time, and set whatever spend limits and usage alerts xAI's console offers so a runaway loop trips a wire instead of a bill. The alert is the cheapest insurance you can buy against the prototype-bug story that opened this chapter — a threshold that pages you at a hundred dollars of unexpected spend turns a catastrophe into a footnote.

Watch the meter, especially early. Monitor your usage as you build and as you grow, the way you would watch any cost that scales with traffic. The bug that runs your call a hundred thousand times does not announce itself; the dashboard is where you catch it. Get in the habit of looking at not just the total but the shape — which model, how much search, which part of the product is spending — because an average hides the one runaway path that is quietly costing more than everything else combined.

Know which meter you are on. API usage bills on its own pay-per-use meter, separate from any consumer Grok subscription (Chapter 20). Do not assume a subscription you pay for as a person covers what your software spends as a program — it generally does not.

The honest summary is the one this whole Part has circled: the Grok API gives you a frontier model and a live line to the present, and both are metered. The structure of that meter — tokens in and out, search and tools on top, rate limits capping your pace — is stable enough to plan around. Every number attached to it is not, and confirming today's figures against xAI's own pricing is the last thing to do before you ship and the first thing to do when the bill looks wrong. Build deliberately, flag the numbers, and let the provider — never this page — tell you what a token costs today.

Part VII: Judgment and Limits

Audience: everyone — and required reading before you lean on Grok for anything that matters.

"Real-time is a superpower and a new way to be wrong, at the same time. Using Grok well is mostly knowing which one you're looking at."

— A working principle

Chapter 23: What Grok Gets Wrong

A book that only told you what Grok does well would be an advertisement, not a guide. The truth is that Grok, like every AI assistant, gets things wrong — sometimes in obvious ways you will catch at a glance, sometimes in subtle ways that slide past a careless read and into your work with your name on them. And Grok has one extra category of failure the others largely do not, precisely because of the thing that makes it special: it reads live, unverified information, and live, unverified information is sometimes false. Knowing the specific ways Grok fails is not pessimism. It is what lets you point your skepticism where it pays off, recognize trouble early, and stay out of the situations where the tool is least reliable. This chapter is the honest inventory.

The Failure Mode That Shapes All the Others: Confidently Wrong

Start with the one that causes the most damage, because it colors how you should read everything Grok produces. Grok does not signal uncertainty the way a careful person does. When a human expert is unsure, they usually tell you — they hedge, they flag the shaky part, they say "I think but I'd check." Grok tends to produce its answer with the same fluent confidence whether it is right or wrong. A correct answer and a subtly broken one arrive in the same assured tone.

This is the deepest reason you cannot outsource your judgment to it. You cannot use the model's confidence as a signal of correctness, because the confidence is roughly constant. **Plausible-looking-but-wrong is the most dangerous thing an assistant produces**, because it sails through a quick skim — it reads well, it sounds sure, and only actually checking reveals the problem. Internalize that one fact and you will read Grok's output with the right reflex; forget it, and the confident wrong answer is exactly the one that gets you, because it is the one you did not think to question.

It is worth understanding *why* the confidence is flat, because the reason tells you the defense. A model like Grok does not "know" things and then report its certainty about them the way a person does; it produces the most plausible continuation of the text, and a fluent, assured sentence is plausible whether or not its content is true. There is no internal "I'm not sure about this" that reliably leaks into the wording. So the confident tone you read is a property of how the model writes, not a readout of how reliable the answer is — and treating it as the latter is the single most common way smart people get burned by these tools. The practical upshot: build your verification habit around *what kind of claim* you are looking at, never around how sure the model sounds, because it sounds equally sure of everything.

The Grok-Specific One: Trusting the Live Firehose

Here is the failure mode that is more pronounced for Grok than for its frozen-in-the-past rivals, and it is the flip side of the entire real-time advantage. When Grok answers from live sources — X posts, fresh web pages, the reaction unfolding right now — it is drawing on information that has not been vetted. The live web is full of rumor, error, motivated spin, coordinated manipulation, and confident nonsense, sitting right next to the truth and often looking identical to it. Grok can faithfully report what people are saying and be completely wrong about what is *true*, because what people are saying is sometimes false.

This bites hardest in exactly the moments you are most tempted to use it: breaking news. In the first hours of a major event, the live picture is at its noisiest — early reports contradict each other, false claims spread fastest, and the correction arrives long after the rumor. Grok asked "what happened" in that window will give you a confident synthesis of a situation that is genuinely not yet known, and it can launder an unverified claim into an authoritative-sounding summary. Real-time sentiment has a related trap: a loud, coordinated, or bot-amplified push on X can look like genuine public opinion, and "what are people saying about X" can return a manufactured consensus as if it were the real one.

Three concrete shapes this failure takes, so you can recognize each in the wild. **The premature verdict:** you ask about an event an hour old, Grok synthesizes the early reports into a clean narrative, and that narrative turns out to be the version that gets corrected by evening — Grok was not wrong about what was being said, it was wrong to make it sound settled. **The manufactured consensus:** you ask how people feel about something and get back a confident "the reaction has been overwhelmingly negative," built on a wave of posts that was, in fact, a coordinated push or a vocal minority that happened to be loud that hour, not the broad sentiment it reads as. **The laundered rumor:** a single unsourced claim circulating on X gets folded into Grok's summary as though it were an established fact, its origin as one anonymous post invisible by the time it reaches you in measured, authoritative prose. In all three, nothing malfunctioned — the model reported the live world accurately and the live world was misleading.

The defense is not to abandon the live edge — it is the best thing Grok has — but to read live answers as *live*: provisional, source-dependent, and worth tracing. Check what the claim is built on (Chapters 9 and 10 are about doing exactly this). Treat breaking-event answers as a first sketch, not a verdict. Distrust unanimity that is too clean. The real-time capability is a superpower and a new attack surface in the same feature, and using it well is mostly knowing which one you are looking at.

Subtle Errors of Fact and Reasoning

Set the live sources aside and Grok still makes the ordinary mistakes every large language model makes. It is strong on the common case and weaker on the precise. It will get the shape of an answer right and a specific detail wrong — a number, a date, a name, a step in a chain of reasoning that does not quite follow. These errors are individually small and collectively the most common real problem, because they survive the "does this look right?" read and only fail when the specific detail actually matters.

Picture the texture of it. You ask Grok to summarize a financial report and the summary is excellent — the right themes, the right emphasis, a clean narrative — except one revenue figure is off by a transposed digit, lifted correctly from the document but then slightly garbled in the retelling. Or you ask it to walk through a calculation and the method is exactly right, the explanation is lucid, and one arithmetic step in the middle is wrong, so the final number is confidently incorrect while every word around it is sound. Or it attributes a real quote to the wrong person, or places a true event in the wrong year. The unifying property is that the error hides inside an answer that is mostly right, which is precisely what makes it slip through — a wholly wrong answer would have set off alarms; a 95-percent-right one does not.

The defense is to verify the load-bearing specifics — the numbers and facts your decision actually rests on — rather than spot-checking the prose for tone. The error lives in the detail you did not check.

Made-Up Specifics That Sound Real

A related trap, inherent to how these models work: when Grok does not know something, it can produce a *plausible-but-nonexistent* specific rather than admitting the gap — a citation to a study that was never published, a quote no one said, a statistic with no source, a fact that fits the pattern of being true without being true. The fluency makes these especially dangerous, because a fabricated detail arrives dressed exactly like a real one. Real-time search mitigates this for current questions — the model can go look rather than guess — but it does not eliminate it, particularly for specifics that fail quietly rather than loudly.

The fabrications cluster in recognizable places, which is useful, because it tells you where to aim your suspicion. **Citations and sources** are the classic one: ask for studies supporting a claim and you may get titles, authors, and journals that look entirely real and do not exist, assembled from the pattern of what such a citation usually looks like. **Quotes** are another: a model will produce a sentence "said" by a named person that captures something they plausibly might have said and that they never actually said. **Precise statistics** are a third: a specific-sounding percentage or dollar figure attached to a claim that has no real number behind it, the precision itself a kind of camouflage. The common thread is that the request asked for a specific the model did not actually have, and the model supplied a plausible one rather than admitting the gap — because supplying a plausible continuation is what it does.

The defense is the oldest one in research: do not repeat a specific you cannot trace to a source, especially a citation, a quote, or a statistic. If Grok gives you one that matters, follow it to the original before you rely on it. A citation you cannot find is not a citation; it is a guess wearing a citation's clothes.

Missing the Context It Could Not Know

Grok acts on the context it has — your prompt, the documents you gave it, the live sources it found. It cannot act on context it does not have: the constraint that lives only in your head, the history behind a decision, the thing everyone in your organization knows and nobody wrote down. So it will confidently answer the question you asked rather than the one you actually meant, or give advice that is generically sound and specifically wrong for your situation. The fix is partly on you — better context in, better answer out (Chapter 7) — and partly just a limit to respect: the model does not know what you did not tell it, and it will rarely stop to say so.

This shows up most painfully in advice. Ask Grok how to handle a difficult conversation, structure a deal, or solve a problem at work, and you will get a competent, generic answer — and competent and generic is exactly what is wrong for a situation defined by its specifics. The history with this particular person, the political reality of your organization, the constraint your boss mentioned once in a hallway, the regulation specific to your industry: none of it is in the model unless you put it there, and the answer it gives you will be confidently blind to all of it. The model will not pause to say "this depends on things you haven't told me." It will simply answer the abstract version of your question as though it were the real one. Respecting this limit means recognizing that the more your situation's correctness depends on details only you hold, the more the model's confident generic answer is a starting point to adapt rather than an answer to follow.

The Edges of the Envelope

Two kinds of task strain Grok the way they strain any model. **Genuinely novel problems** — the ones with no established pattern, demanding real invention or deep specialized expertise — are where it is weakest, because it works largely by competent recombination of what it has seen, and true novelty is what it has not seen. And **highly specialized or high-stakes domains** — law, medicine, finance, anything where a wrong answer is expensive and the right one requires a credentialed professional — are places to use Grok as a starting point and a thinking aid, never as the final authority. It can help you understand and prepare; it cannot take responsibility, and it does not carry the liability when it is wrong. You do.

The high-stakes case deserves the bluntest version, because the temptation to lean on the tool is strongest exactly where the cost of its being wrong is highest. Grok can be genuinely useful for understanding a legal concept before you talk to a lawyer, for framing questions to bring to a doctor, for getting oriented in an unfamiliar financial decision. What it cannot do is be the lawyer, the doctor, or the advisor — not because it is incapable of producing relevant words, but because being the professional means carrying the accountability and the duty of care that no model carries. When Grok is wrong about your medication, your contract, or your taxes, the consequence lands on you, not on it. That asymmetry — the model takes none of the risk and you take all of it — is the real reason these domains demand a credentialed human in the loop, and it does not soften just because the model's answer was fluent and sounded expert.

What This Inventory Is For

The point of cataloging the failures is not to scare you off a genuinely capable tool. It is to make you good at operating it, because each failure maps to a defense you now have: skepticism toward confident output, source-tracing for live answers, verification of load-bearing specifics, following fabricated-sounding citations to ground, good context in, and human ownership of the novel and the high-stakes. The failures are patterned, and patterns can be defended against.

The right posture is neither credulous nor cynical. Grok is genuinely capable and genuinely fallible, and using it well means holding both at once — leaning on the enormous amount it does well while keeping the specific, informed skepticism that catches the specific, knowable ways it fails. That is not distrust of the tool. It is competence with it. The next chapter turns from the errors Grok makes by accident to the most debated thing about it on purpose: its tone, its politics, and the "less filtered" reputation that is either its appeal or its liability, depending on who you ask.

Chapter 24: Bias, Tone, and the "Less Filtered" Question

This is the chapter people skip to, because the most-repeated thing about Grok is not its context window or its real-time search — it is its reputation for being blunter, edgier, and "less filtered" than the other assistants. That reputation is the source of both its appeal to some users and its notoriety with others, and it is impossible to use Grok thoughtfully without forming a clear-eyed view of it. So I am going to treat the question directly and without either of the two easy postures: the cheerleading that says Grok is the only honest AI in a censored field, and the panic that says it is uniquely dangerous. Both are too simple. The reality is more useful, and more worth your attention.

Where the "Less Filtered" Trait Comes From

Recall the founding story from Chapter 2. xAI was started on the explicit thesis that mainstream AI had become too cautious — too quick to refuse, too hedged, too willing to wrap every answer in disclaimers and decline anything contentious. Grok was built as a deliberate reaction to that: more willing to engage with edgy or sensitive topics, more willing to give a direct opinion, more willing to have a voice with some personality and some bite. The "less filtered" quality is not an accident or a bug. It is a stated design goal, and the company has positioned "maximally truth-seeking" and less-hedged engagement as features, not liabilities.

It is worth being precise about what that does and does not mean, because the marketing language ("truth-seeking," "based," "anti-woke" in some of the discourse around it) can obscure the actual behavior. In practice, "less filtered" means a few concrete things: Grok will more often answer a question other assistants decline; it will more readily offer a blunt take rather than a carefully balanced both-sides summary; and its register is livelier, sometimes jokier, sometimes sharper. That is the trait. Whether it is a virtue or a hazard depends entirely on what you are using it for.

A point that often gets lost in the heat around this: the "filter" on any AI assistant is not one dial but several distinct things bundled under one word, and Grok loosens some of them and not others. There is the question of what topics the model will *engage with* versus refuse. There is the question of how much it *hedges* — whether it commits to a position or surveys every side. There is the question of its *register* — how formal, how playful, how willing to crack a joke or take a swing. And there is the question of its *safety restraints* — the hard limits meant to keep genuinely harmful content from coming out. These are not the same thing, and conflating them is where most of the confused arguments about Grok come from. A model can be loosened on register and hedging — livelier, more opinionated — while keeping firm safety restraints, or it can be loosened on the restraints themselves, which is a very different and more consequential choice. When you read "less filtered," it pays to ask *which* filter, because the lively voice that makes Grok pleasant to brainstorm with and the loosened restraint that has occasionally produced real trouble are not automatically the same loosening.

Where Directness Is Genuinely Useful

Let me give the trait its honest due, because there are real situations where the other assistants' caution is a cost and Grok's willingness is a benefit.

When you want a **straight answer rather than a hedge**, Grok is often more useful. Ask a pointed question and you are more likely to get an actual position you can react to, instead of a noncommittal survey of every viewpoint that leaves you exactly where you started. For brainstorming, for pressure-testing an argument, for getting an opinion you can push against, directness has real value.

Picture the difference in a concrete moment. You have written an argument and you want it stress-tested. Ask a heavily hedged assistant "is this argument any good?" and you may get a balanced appraisal that finds something nice to say about every part and never quite tells you where it is weak — polite, and nearly useless for your purpose. Ask Grok the same thing and you are more likely to get "the second point doesn't hold up, here's why" — a position you can actually work against. The value is not that Grok's verdict is automatically right; it is that a committed verdict is a thing you can push on, sharpen, or reject, whereas a survey of all possible views gives you nothing to grip. For the specific job of having your thinking challenged rather than soothed, the willingness to commit is the feature.

When a topic is **legitimate but sensitive** — health questions adults are entitled to ask, controversial history, difficult subject matter for research or writing — an assistant that engages rather than reflexively refuses can simply be more helpful, and less condescending about your ability to handle information.

And there is a **personality dimension** that some users genuinely prefer: an assistant that is a bit more human, a bit less corporate, a bit more willing to be funny or opinionated. That is a matter of taste, not capability, but it is a real reason some people choose Grok, and it is a legitimate one.

Where It Has Caused Real Trouble

Now the other side, stated just as plainly, because a guide that sold you the upside and hid the downside would be exactly the kind of dishonest writing this series tries not to be.

Grok's looser guardrails have, on multiple well-documented occasions, produced outputs that crossed from "direct" into genuinely harmful — offensive content, hateful or extreme statements, and material that xAI itself had to publicly walk back, apologize for, and correct. Some of these episodes were traced to deliberate changes in how the model was instructed to behave; some were the predictable result of loosening the restraints that, on other systems, catch this kind of thing before it reaches a user. Either way, they are not hypothetical risks. They happened, in public, and they are part of the honest record of what this tool is.

The pattern across these episodes is the instructive part, more than any single incident. What recurs is that a change intended to make the model more direct, more willing, less hedged — applied at the level of how the model is instructed to behave — produces consequences beyond what was intended, because the same loosening that removes an unnecessary refusal can also remove a necessary restraint. The dial that says "engage more readily" does not distinguish neatly between "engage with the sensitive question an adult is entitled to ask" and "engage with the thing that should have been declined." That is not an argument that the tuning cannot be done well; it is an observation that it is genuinely hard, that the failures are the kind that surface in public rather than in testing, and that a tool tuned toward willingness will, by the nature of the tuning, fail in the direction of saying too much rather than too little. Knowing that tells you where this tool's characteristic mistakes will land.

There is also a **political dimension** that you cannot pretend away. xAI's founder is one of the most politically prominent and polarizing figures in the world, with very public views, and Grok has at times reflected — or been steered toward — positions that align with them, in ways that have drawn pointed criticism and raised a fair question: when an AI is described as "truth-seeking," whose account of the truth is being encoded? Every AI assistant carries the fingerprints of the people and choices that shaped it; none is a neutral oracle. Grok's are simply more visible and more contested than most, partly by design and partly because of who is behind it.

It is worth being even-handed about this rather than singling Grok out unfairly. The honest version of the point is not "Grok is biased and the others are neutral" — none of them is neutral. Every one of these systems was shaped by choices about what data to train on, what to reward, what to refuse, what voice to adopt, and those choices encode a perspective whether or not anyone intended a political one. What is distinctive about Grok is not that it has a perspective but that the perspective is *more legible*: tied to a named, vocal, polarizing owner, sometimes steered in visible ways, and marketed under a "truth-seeking" banner that invites exactly the question of whose truth. The right takeaway is not "distrust Grok and trust the others"; it is "read every one of these tools as a thing with a point of view, and read Grok's with the added awareness that its point of view is unusually traceable to a particular person and a particular project."

The uncomfortable conclusion is the one from Chapter 2: the trait that makes Grok willing to engage where others refuse, and the trait that has periodically produced trouble, are the *same trait*. You do not get the directness without also getting the looser restraints that occasionally fail. Strength and liability share a root, and any honest assessment has to hold both.

How to Actually Think About It

So what do you do with a tool like this? A few principles that have served me.

Treat directness as an input, not a verdict. A blunt, confident take from Grok is a useful thing to react to — and exactly the thing not to mistake for established fact. The willingness to give an opinion does not make the opinion correct. Use it to sharpen your own thinking, not to replace it.

Know that "less filtered" does not mean "more true." This is the crucial misread to avoid. Fewer guardrails means the model will *say* more, not that what it says is more accurate. An assistant that confidently states something another assistant declined to state is not thereby more honest; it may just be less careful. The "truth-seeking" branding is a marketing claim, not a property you should assume.

Match the tool to the context. For private brainstorming and personal use where you want a direct interlocutor, the trait is often an asset. For anything public-facing, professional, regulated, or representing someone other than yourself, the looser guardrails are a genuine risk to weigh — which is the entire subject of Chapter 26. The same bluntness that is a gift in your private notebook is a liability the moment its output carries your employer's name, because the directness that you can contextualize for yourself, a customer or a regulator will read at face value.

Keep your own bias-awareness on. Grok has a perspective, as every model does. Knowing that, read its answers on contested topics the way you would read a sharp, opinionated columnist: worth engaging, not to be swallowed, and always worth asking "what would someone who disagrees say" — a question you can put to Grok itself. That last move is genuinely useful: a model tuned to give a direct take will, if you ask it to, also give you the strongest version of the opposing case, and reading the two against each other does more to sharpen your judgment than either alone.

The Even-Handed Bottom Line

Grok's tone is its most distinctive and most divisive trait, and the honest verdict refuses both caricatures. It is not the one brave truth-teller in a field of censored cowards, and it is not a uniquely reckless machine you should be afraid of. It is an assistant deliberately built with a livelier voice and looser restraints, which makes it genuinely more useful for some tasks and genuinely riskier for others, and which has the public track record — good and bad — to prove both halves of that sentence. Use it knowing exactly which trait you are leaning on, and you will get the benefit of the directness while keeping your guard up for the times it tips over.

That guard matters most in two specific places: what happens to your data when you use a tool this tied to a live social platform, and what it means to bring an assistant with these traits into a workplace. Those are the last two chapters, and they are where the abstract considerations of this one turn into concrete decisions.

Chapter 25: Privacy and Your X Data

Every AI assistant raises the question of what happens to the things you type into it. Grok raises a sharper version of that question than most, for one structural reason: it is made by the company that owns X, and it is wired into the same platform where you may have spent years posting. That tie is the source of Grok's signature strength, and it is also the source of a privacy story you should understand before you lean on the tool — both for what you put *into* Grok and for what X already has *about* you. This chapter lays out the considerations honestly. As always with specifics, the authoritative source is xAI's and X's current privacy policies and settings, which change; what follows is the durable shape and the questions to ask, not a substitute for reading today's terms.

Two Different Privacy Questions

It helps to separate two things that get tangled together, because they have different answers.

The first question is about your conversations with Grok — the prompts you type, the documents and images you upload, the chats you have. Like most AI providers, xAI may use this interaction data to operate and improve the service, which can include having it inform model training and, in some cases, being reviewed by humans for quality and safety. The practical upshot is the same one that holds for every assistant in this series: **treat what you type into Grok as something that may not stay private to you.** That is the baseline to internalize before anything provider-specific.

The second question is about your X data — your posts, your activity, the public record you have built on the platform. This is the part that is distinctive to Grok, and it deserves its own treatment, because it is where the assistant's deep integration with a social network creates a privacy dimension the standalone assistants simply do not have.

The reason it pays to keep these two questions apart is that they pull in different directions, and a person who lumps them together tends to act on the wrong one. Plenty of users worry intensely about the first — they hesitate over typing a half-finished idea into a chat box — while never once looking at the second, which is the setting that actually governs years of accumulated posting history. Others do the reverse: they toggle one X switch, feel protected, and then paste a confidential contract into Grok the next afternoon. The two questions need two different habits, and the rest of this chapter builds each one.

Your X Posts and Model Training

Because xAI and X share an owner, the enormous corpus of public posts on X is available to xAI as training material in a way an outside company's data is not — and that has included, at various points, using X users' posts and interactions to train Grok. For most people this concerns *public* content they chose to post to a public platform, which changes the calculus somewhat: a public tweet was never private. But many users were surprised to learn their activity was feeding an AI model, and the arrangement has been contentious, has drawn regulatory attention in some jurisdictions, and has been the subject of changing settings and defaults.

The concrete thing to do is this: **find the current data-sharing setting in your X account and decide deliberately whether you want your activity used to train Grok.** There has historically been a control governing this, and its existence, location, and default state have shifted over time — which is precisely why you should look at it yourself rather than assume. If you are privacy-conscious, this is the single most important Grok-related setting you own, and it lives in your X account, not in Grok. Verify where it is today and set it the way you want.

Here is the practical walkthrough, with the caveat that the exact menu path will have drifted by the time you read it. The setting has historically lived under the web client's **Settings and privacy → Privacy and safety → Data sharing and personalization**, in a section about Grok or about sharing data with xAI, presented as a checkbox you can clear. Two things are worth knowing about it. First, it has at times been **opt-out rather than opt-in** — meaning the default was "on," and doing nothing was a choice to participate, not to abstain. If you never touched it, do not assume you were excluded. Second, clearing the box governs *future* use; it is not a magic eraser for training that has already happened. A model that has already learned from your past public posts does not unlearn them because you changed a setting today. The setting is a forward-looking valve, and the value of closing it is real but bounded by when you close it.

There is a subtler point hiding in the word *interactions*. The training material is not only the posts you compose. It can extend to the texture of your activity — what you like, what you reply to, what you linger on — which is a richer and more revealing record than your posts alone. Someone who rarely posts but reads voraciously still leaves a substantial trail. If the training question matters to you, the thing you are deciding about is your whole footprint on the platform, not just the part of it that is visible to others.

A Word on Jurisdiction

Where you live changes your options here, and it is worth a paragraph because the differences are concrete rather than theoretical. In the European Union and the United Kingdom, data-protection law (the GDPR and its UK equivalent) gives individuals rights over personal data — including, in many readings, a basis to object to certain processing — and regulators there have already pressed xAI over exactly the practice of training on users' posts, which is part of why the relevant settings and defaults have moved. If you are an EU or UK user, you generally have stronger levers and clearer recourse than a user elsewhere, and the platform is obliged to give you a real way to say no. In the United States, the baseline is weaker and more patchwork; a handful of states have their own privacy statutes, but there is no single federal equivalent, and the default tends to favor the provider's latitude. None of this changes the basic hygiene advice. It changes how much standing you have to insist, and it is the reason the same setting can appear, disappear, change its default, or sprout new options depending on where the platform thinks you are.

What This Means in Practice

Pull the two questions together and a few practical rules fall out, none of them exotic — they are the same digital hygiene that serves you everywhere, applied to a tool with an unusually direct line to a social platform.

Do not put genuinely sensitive information into Grok. Trade secrets, confidential business material, other people's personal data, passwords and credentials, anything regulated or legally protected — keep it out, the same as you would with any consumer AI assistant. If a leak of it would matter, it does not belong in a consumer chat box.

It helps to have a quick internal test for "genuinely sensitive," because the category is easy to wave at and hard to apply in the moment. Try this one: *would I be comfortable if this exact text turned up, attributed to me, somewhere I did not put it?* A recipe idea, a draft of a blog post, a question about how a Python loop works — fine; the honest answer is yes, who cares. A client's unreleased financials, a colleague's medical situation, the password to your email, the clause your lawyer told you to keep quiet — the honest answer is no, and that "no" is your signal to close the box and do the work another way. The test is useful precisely because it does not require you to know xAI's retention schedule. It only requires you to know your own exposure, which you always do.

Remember that the assistant and the platform are connected. When you use Grok inside X, or ask it about your own posts, you are operating inside an integrated system owned by one company. That is convenient and it is also worth holding in mind: the line between "my private assistant" and "the social platform I post on" is thinner here than it is with an assistant from a company that does not also run your social feed. A concrete illustration: ask a standalone assistant "what have I been posting about lately and how is it landing," and it has no idea who you are. Ask Grok inside X the same thing and it may be able to answer, because the account, the posts, and the assistant all sit under one roof. That is a genuinely useful capability — and it is exactly the kind of capability that should make you think about how much of your own record you want the tool reaching into by default.

Check your X privacy settings, not just Grok's. Because the data story spans both products, the controls you care about are split across both. The Grok-specific setting (whether your X activity trains the model) lives on the X side. Spend ten minutes in both settings panels and you will know far more than most users about where your data goes. While you are there, it is worth glancing at the adjacent controls that shape what Grok can even see — the visibility of your posts, whether your account is protected, who can see your activity — because a tightly locked-down account exposes less to *every* downstream use, training included.

Assume public is public. Anything you have posted publicly on X is, by definition, already out in the world and reasonably treated as fair game for analysis — by Grok and by everyone else. The privacy decisions worth agonizing over are about your *private* information and your *future* sharing, not the public record you already created. This cuts the other way too, and usefully: it means the highest-value privacy move you can make is not retroactive cleanup of old posts but **a small change in what you commit to the record going forward.** You cannot un-post a decade of tweets in any way that matters; you can absolutely decide that the half-formed political rant or the detail about your kid's school is one you keep off a public platform from here on.

Deleting, Exporting, and the Limits of Both

People reasonably ask: can I just delete my chats, or download my data, and be done? Both are worth doing and neither is a complete answer, so it is worth being precise about what each buys you.

Grok and X both offer some ability to **delete your conversation history**, and you should use it as ordinary hygiene — there is no reason to keep a year of accumulated chats sitting in an account. But deleting a chat from your view is not the same as guaranteeing it has been purged from every backup, log, and any training corpus it may already have informed. Deletion controls the *interface* and your *account*; they make far weaker promises about derived data and historical processing. Treat deletion as good housekeeping, not as a guarantee of erasure.

Likewise, both platforms typically let you **export an archive** of your data, which is genuinely useful for seeing what is held about you and is in some jurisdictions a legal right. An export is a snapshot of what the account holds; it is not a list of everywhere your data has flowed or what a model has internalized from it. Use the export to inform your decisions — it is illuminating to actually see the record — without mistaking it for the full accounting of where your information lives.

Consumer Versus Enterprise

One distinction matters enough to flag on its own, because it changes the answer entirely. The privacy posture of consumer Grok — the free tier, the personal subscriptions, the version inside X — is governed by consumer terms that generally allow the provider broad latitude to use your data to improve the service. The privacy posture of **enterprise and business offerings**, where they exist, is typically very different: stronger contractual data-handling guarantees, commitments not to train on your data, retention controls, and the compliance features organizations require. If you are evaluating Grok for a company rather than for yourself, do not reason from the consumer terms — the business terms are a separate document and a separate conversation, and they are where the data guarantees that actually matter to an organization get made.

The same split exists, in miniature, for developers using the API. Data sent through a paid developer endpoint is often governed by terms closer to the business posture than the consumer one — frequently with explicit statements about whether inputs and outputs are used for training. If you are building on Grok rather than chatting with it, read the developer terms specifically (Part VI), because "what happens to my data" can have a materially different answer at the API than it does in the consumer app, and you should know which set of promises you are operating under. Chapter 26 picks the workplace side of this up in detail.

The Honest Summary

Grok's privacy story is not uniquely alarming, but it is uniquely *shaped* by the thing that makes Grok distinctive. The conversation-data question is the same one every assistant raises, and the same answer applies: assume it may be used, and keep the genuinely sensitive things out. The X-data question is Grok's own, and it has a concrete remedy you have already met: the training control in your X settings, set on purpose rather than left on its default. Do those two things — guard what you type, and own the setting that governs your posts — and you have handled the great majority of what there is to handle. The rest is the ordinary judgment of not putting into any consumer tool what you could not stand to see leave your control.

That judgment becomes much higher-stakes the moment you are acting on behalf of an employer rather than yourself, because then the data is not only yours to risk and the looser-guardrails question from Chapter 24 meets real organizational consequences. Bringing Grok to work responsibly is the final chapter, and it is where everything in Part VII comes together into a set of decisions you can actually make.

Chapter 26: Responsible Use at Work

Everything in this book changes character the moment you are using Grok on behalf of an employer rather than for yourself. At home, the consequences of a wrong answer, an overshared detail, or an off-color output land on you. At work, they land on your colleagues, your customers, your organization's reputation, and sometimes its legal exposure — and the looser, livelier traits that make Grok appealing in private become liabilities you have to manage in public. This final chapter pulls the threads of Part VII together into the practical question every professional reader actually faces: how do I get the real value of this tool at work without creating a problem? The answer is not "don't." It is "do, deliberately, inside a few boundaries."

Start With What Not to Feed It

The first rule is the simplest and the most violated: **do not put confidential or regulated information into a consumer AI tool.** Chapter 25's rule carries straight over and only hardens here — none of it, from customer data to trade secrets to unreleased financials to proprietary code to anything under contract or regulation, belongs in a consumer Grok session. The conversation data may be used to improve the service; the genuinely sensitive thing you typed may not stay yours. This is not specific to Grok — it is true of every consumer assistant in this series — but Grok's tight integration with a social platform makes the instinct to keep the sensitive things out worth holding especially firmly.

It is worth making this concrete, because "confidential information" is a phrase people nod at and then violate within the hour, usually without noticing. Consider the ways it actually happens at work. A salesperson pastes a prospect's full contract into Grok to get a plain-English summary before a call — and has just put a counterparty's confidential terms into a consumer tool. A developer pastes a stack trace to debug it, and the stack trace happens to include an internal hostname, an API key, and a snippet of customer data in the payload. A manager drafts a sensitive performance review by describing a named employee's struggles and asking Grok to "make this more diplomatic" — and has now disclosed personnel information about a real person. A finance analyst pastes a row from a spreadsheet to check a formula, and the row is next quarter's unannounced revenue. None of these people set out to leak anything. Each one was just trying to do their job faster, which is exactly why the failure is so common: the convenience and the breach are the same motion.

The defense is not vigilance in the moment — vigilance fails precisely when you are busy and the tool is helpful. The defense is a couple of habits that work even when you are not thinking. **Redact before you paste:** strip names, keys, account numbers, and identifiers and replace them with placeholders, because Grok almost never needs the real ones to help you. Ask it to fix the *structure* of a contract clause without the parties' names; ask it to debug the *shape* of an error without the live credentials in it. And **build the reflex of asking "whose information is this?"** before sensitive text goes in. If the answer is anyone but you — a customer, a colleague, a counterparty, the company itself — that is the prompt to slow down.

The way around this limit, where it exists, is not to ignore it but to use the right door: an enterprise or business offering with contractual data-handling guarantees (Chapter 25), procured and vetted by the people whose job that is. If your organization wants to use Grok on real internal data, that is the path — not individuals quietly pasting confidential material into a personal account because it was convenient. The difference between the two doors is not cosmetic. The enterprise door comes with a contract that says, in writing, what the provider may and may not do with your data — often including a commitment not to train on it, defined retention, and the audit and access controls a compliance team can actually point to. The consumer door comes with terms written to give the provider latitude. Same model, very different promises, and at work it is the promises that govern your exposure.

Respect the Looser-Guardrails Risk in Public Contexts

Chapter 24 was honest about Grok's "less filtered" trait: genuinely useful for some work, genuinely risky for other work. At work, the dividing line is usually whether the output is going to be seen by, or represent, anyone other than you.

For **private, internal thinking** — drafting you will heavily revise, brainstorming, pressure-testing an argument, summarizing for your own understanding — the directness is mostly an asset, and the risk is low because you are the only audience and the editor.

For anything **public-facing or representational** — customer communications, published content, anything with the company's name on it, anything a client or regulator or journalist might see — the looser guardrails are a real hazard. An assistant that will more readily produce a blunt, edgy, or unvetted statement is an assistant whose raw output you must never ship unreviewed in a context where tone and accuracy carry institutional consequences. The rule is simple: the more public the destination, the more thoroughly a human owns the final word. Grok can draft; a person decides what actually goes out.

A worked example makes the dividing line easier to feel. Suppose you run social media for a brand and you want help responding to a wave of angry posts about a product delay. Asking Grok to *analyze* the complaints — cluster them, tell you which themes recur, gauge the temperature of the conversation in real time — plays directly to its strengths, and nothing about that touches the public. The moment you ask it to *write the public reply*, the calculus flips. The trait that makes Grok pleasant to argue with in private is precisely the trait that produces a reply with too much edge, a defensive jab, or a confident claim about the delay's cause that turns out to be wrong — and now it has your brand's name on it and screenshots itself. The responsible workflow uses Grok for the half it is good at (read the room, draft a rough first pass) and puts a human firmly in front of the half where institutional voice and accuracy are on the line. Drafting is not publishing, and at work the gap between them is where your judgment earns its keep.

Verify Before It Becomes a Decision

The confidently-wrong problem and the noisy-live-sources problem from Chapter 23 are not academic at work — they are how a bad number ends up in a board deck or a false "breaking" claim ends up in a customer email. The discipline is proportional: the verification you owe an output should scale with the cost of it being wrong.

For low-stakes internal use, a sensible read is enough. For anything that informs a real decision, goes to a customer, or gets published, the load-bearing facts — the numbers, the claims, the citations, the live-sourced "what's happening" syntheses — must be independently confirmed before they harden into action. This is especially true of Grok's real-time answers, which are at their noisiest exactly when they are most tempting: in the fast-moving moment. "Grok said so" is not a source you can stand behind in a professional setting. The original it drew from is.

The phrase to hold onto is *load-bearing*. Not every fact in a draft needs the same scrutiny; the ones that do are the ones a decision rests on. If a memo casually mentions that a market is large and growing, that is texture and a loose read is fine. If the memo recommends a hundred-thousand-dollar bet *because* the market is growing at a specific rate Grok reported, that rate is load-bearing and you trace it to a primary source before anyone acts on it. The skill is learning to feel the difference inside your own output — to notice which sentences are doing real work and check those hardest. A useful tell: Grok's real-time answers are most seductive and least settled in the first hour after an event, when the live sources it is reading are themselves rumor and guess. The closer your question is to "what just happened," the more the answer is a starting point for verification rather than a finding you can quote. Treat a fresh real-time synthesis the way a careful editor treats a wire report from a breaking story: useful, fast, and not yet confirmed.

Be Transparent About Using It

Norms here are still forming and they vary by organization and profession, but the safe default is openness rather than concealment. If AI-assisted work would matter to the people receiving it — a client expecting your original analysis, a publication with disclosure rules, a colleague who needs to know a draft was machine-generated before they rely on it — say so. Quietly passing off unreviewed AI output as your own careful work is how trust gets broken when it inevitably surfaces, and it surfaces more often than people expect. Used openly as the productivity tool it is, Grok is unremarkable; used secretly as a substitute for the judgment you are being paid for, it is a problem waiting to happen.

Transparency does not mean stapling a disclaimer to every email; it means matching disclosure to what the recipient would reasonably want to know. Three rough tiers help. When AI is just a **tool in your process** — it helped you draft, summarize, or check something, and you reviewed and own the result — no announcement is usually needed, the same way you do not disclose using a spreadsheet or a spell-checker. When AI **materially shaped a deliverable** someone is relying on as your work — a client expecting your independent analysis, a report whose conclusions came substantially from the model — a plain note is the honest move. And when there is a **rule or expectation** in play — a journal's disclosure policy, a court's, an academic standard, a client contract — you follow it to the letter, because there the disclosure is not a courtesy but an obligation. The test underneath all three is the same: *would this person feel misled if they later learned how the work was made?* If yes, you say so now, while it is a normal note rather than later, when it is a discovered concealment.

Work Within Your Organization's Rules — and Help Make Good Ones

Many organizations now have AI-use policies, and some specifically address which tools are sanctioned, what data may go into them, and what review is required. **Know yours before you adopt Grok for real work.** A tool being capable does not make it approved, and "I didn't know we weren't allowed to use that" is a poor position to be in after the fact. If your organization has procured an enterprise arrangement, use that rather than a personal account. If it has restricted a category of tools, respect the restriction or change it through the right channel.

There is also a layer above your own organization's rules that is starting to bite, and a professional reader should at least know it is there. In the European Union, the **EU AI Act** is now a live regulatory regime rather than a proposal, and it phases in over time: its provisions on AI literacy and on certain prohibited practices (Article 4 and the related early obligations) became applicable on **2 February 2025**, and broader obligations for general-purpose and higher-risk systems continue to phase in, with major milestones around **2 August 2026**. You do not need to be a lawyer to take the practical lesson, which is that "is this tool allowed and on what terms" is increasingly a question with a *legal* answer in some jurisdictions, not only an internal-policy one — and that the answer can depend on how the tool is used (a low-stakes drafting aid and a system making consequential decisions about people are treated very differently). If your work touches regulated decisions or EU users, the relevant rules are someone's job to track, and the responsible move is to find out who and stay inside what they tell you.

And if you are one of the people who *makes* those rules, a few principles drawn from this whole book: write policy around the durable risks (sensitive-data handling, verification of high-stakes output, public-facing review, disclosure) rather than around specific feature names that will be obsolete in a quarter; distinguish the genuinely useful low-risk uses you want to enable from the high-risk ones you need to govern, rather than banning everything and driving usage underground; and treat Grok's particular traits — the real-time strength, the looser guardrails, the social-platform data tie — as real and specific factors in where it fits, not reasons for blanket approval or blanket prohibition. The blanket-ban instinct is worth resisting specifically: a policy that forbids a tool people find genuinely useful does not stop the usage, it just moves it onto personal accounts and personal phones where you have no contract, no audit trail, and no idea what data is going where. A policy that says "here is the sanctioned way to do this, and here are the three things you must never put in" gets far better compliance than one that says "no," because it works with the reason people reach for the tool instead of against it.

Where Grok Earns Its Place at Work

Net it out and the responsible position is neither reckless adoption nor reflexive avoidance. Grok earns its place at work in the spots its strengths are real and its risks are contained: real-time research and monitoring where being current is the whole value (and the sources get checked), internal drafting and analysis you will review and own, and the everyday productivity work every capable assistant does well. It earns *caution* exactly where Part VII has pointed all along: confidential data, public-facing output, high-stakes decisions, and any context where the looser guardrails could embarrass or expose the organization.

The professionals who get this right are not the ones who use Grok the most or the least. They are the ones who use it with intent — who know what it is genuinely good at, keep their guard up where it is weak, never let its confidence substitute for their verification, and never let its convenience override their organization's rules or their own accountability. That accountability is the thing that does not transfer. Grok can do an astonishing amount of the work; the responsibility for what you do with it stays exactly where it started, with you. Hold onto that, and everything else in this book becomes usable rather than hazardous.

That is the honest close of the argument. What remains is the short reflection that follows — on what it actually means to work with an assistant that lives in the present — and then the reference material for the days you need a specific answer rather than the reasoning behind it.

Epilogue: The Real-Time Advantage

This book opened on a Tuesday-morning question that two assistants could not answer and one could, and it is worth ending where it began, because that small moment contains the whole argument. The specifics in these pages will age — the model will not be Grok 4.3 forever, the prices will move, the features will gain and shed capabilities, and the screenshots will look quaint. That is the nature of writing about something this alive. But the shape of the thing does not age, and the shape is what I want to leave you with.

For most of the short history of AI assistants, the tools lived in the past. They were trained up to a date and then frozen, brilliant about everything that happened before the cutoff and silent about everything after. We learned to work around that silence — to stop asking about anything recent, the way you learn not to ask a reference book about this morning. The blind spot became so normal we stopped noticing it was a blind spot at all.

Grok's contribution, more than any single feature, is to treat the present as a thing an assistant can reach rather than a thing it has to apologize for. That sounds like a small adjustment and it is not. An assistant that knows what happened this morning, that can read the live reaction to an event as it unfolds, that can tell you what is being said right now instead of what was true at training time, is doing a different job than its frozen predecessors — not a better-at-everything job, but a genuinely different one, with the present folded into the set of questions you are allowed to ask. Once you have used a tool that lives in the present, the old limitation stops feeling like the natural order and starts feeling like what it always was: a constraint, now lifted.

That is the durable idea, and it is why the real-time material sat at the center of this book rather than off to the side. DeepSearch, the live tie to X, the willingness to go and look instead of reconstruct from memory — these are not a feature list. They are one capability, seen from different angles, and that capability is the reason Grok exists as its own thing rather than another entry in a crowded field.

What Actually Changes in How You Work

It is worth being concrete about what this difference looks like once the novelty wears off, because the lasting change is not the dramatic "breaking news" moment — it is quieter than that, and it accrues. For years, working with an AI assistant trained you to pre-filter your own questions. Before you asked, some part of you ran a check: *is this too recent for the tool to know?* If the answer was yes, you did not bother; you opened a browser and did the legwork yourself, and over time you stopped even forming the recent questions, the way you stop reaching for a light switch in a room you know has none. That self-censoring became invisible because it was constant.

The real shift with a present-aware assistant is that the filter comes off. You start asking the recent questions again — not because any single answer is miraculous, but because the category of question you had quietly retired is available once more. "Did that actually ship?" "What's the reaction to the announcement an hour ago?" "Is this rumor holding up or falling apart?" These were questions you used to route around your assistant entirely. Now they are questions you route *through* it. The work you used to do around the tool becomes the thing you ask the tool to do, and the saved motions add up across a week far more than the occasional spectacular result does. That is the change worth internalizing: not that the assistant can do one amazing thing, but that an entire class of question came back into play.

But the same honesty that ran through every chapter has to run through the close, because the real-time advantage comes bound to its shadow, and pretending otherwise would betray the whole spirit of the book. **Living in the present means living in the noise.** The live web that lets Grok answer "what's happening right now" is the same live web full of rumor, manipulation, and confident falsehood. The directness that makes Grok useful is the same directness that has, on the record, tipped into trouble. Every one of Grok's distinctive strengths arrives fastened to a distinctive risk. That is not a flaw in the tool so much as the deal it offers, and using it well is mostly a matter of taking the deal with your eyes open.

It is worth naming the pattern plainly, because it has been the spine of this book whether or not it announced itself. The real-time reach that is Grok's great strength is the same reach into unverified, fast-moving sources that makes its current-events answers noisiest exactly when they are freshest (Chapters 9 through 11, and the verification discipline of Chapter 23). The willingness to give a direct, unhedged take that makes it a good interlocutor is the same loosened restraint that has produced public trouble (Chapter 24). The deep integration with X that lets it know what the world is saying is the same integration that gives it an unusual line to your own data (Chapter 25). Three strengths, three shadows, and in every case the strength and the shadow are not two features that happen to coexist — they are one design choice seen from two sides. You do not get to keep the upside and discard the downside, because they are the same thing. That is the deal, and a reader who has understood that one structural fact has understood the most important thing this book had to say.

Which is Where You Come In

And it is where you stay. The thing this book kept returning to, chapter after chapter, is that the assistant's capability does not reduce the need for your judgment — it raises it. A tool that confidently reports the present is a tool whose reports you have to be able to weigh. A tool that will give you a blunt take is a tool whose takes you have to be able to evaluate. A tool that reads a noisy live source is a tool whose sources you have to be willing to check. The more the assistant can do, the more it matters that you can tell when it is right, and that skill — knowing what good looks like, sensing when an answer is too clean, tracing a claim to its origin, holding accountability for what you do with what it tells you — does not automate away. It is exactly what you use to check the automation, and it grows more valuable as the automation grows more capable.

There is a reassuring corollary buried in that, and it is worth saying out loud as we close. People worry, reasonably, that tools like Grok make human judgment obsolete — that if the machine can research, draft, analyze, and report, what is left for the person? The honest answer this book has argued for is the opposite. A more capable assistant does not retire your judgment; it relocates it. The work moves up a level, from doing the legwork to discerning whether the legwork is sound, from finding the facts to weighing them, from producing the draft to deciding what is fit to send. That higher-level work — knowing what matters, knowing when something is off, owning the outcome — is the part that was always the valuable part. The tool takes over the part that was tedious and leaves you the part that was the point. That is not a threat. It is, if you let it be, a promotion.

So the closing advice is simple, and it mirrors the advice this whole series gives about every powerful tool. Reach for Grok where its strengths are real: when the present is the question, when you want a direct interlocutor rather than a hedge, when being current is the whole value of the answer. Keep your guard up where its risks are real: the noisy live source, the confident error, the looser guardrail, the data tie to a platform you also post on. And never let its fluency stand in for your verification or its convenience override your responsibility, because those are the two trades that turn a genuinely remarkable tool into a genuinely avoidable problem.

The present is the one place a frozen model can never go, and Grok was built to live there. That is a real and lasting advantage — and like every real advantage, it is worth exactly as much as the judgment you bring to using it. The tool supplies the present. What you make of it is, as it has always been, up to you.

What follows is reference — the model and pricing snapshot, the API quickstart, the glossary, the resources for staying current — there for the day you need a specific answer rather than the argument behind it. The argument, by now, you have.

Appendix A: Model and Pricing Reference

This is the most perishable page in the book, and I want to say so plainly before you read a single number. Model names, version strings, context sizes, token rates, tier names, and what each tier unlocks are *exactly* the details that move on a scale of weeks to months at xAI. What follows is a mid-2026 snapshot, offered as a reference point so you have a shape to reason against — but **the only authoritative source is xAI's own current model and pricing documentation, and you should verify everything here against it before making any decision that turns on a figure.** Treat this appendix as "roughly the shape of things as of writing," not as current fact.

The Model Family (Mid-2026)

Grok is powered by the **Grok 4.x generation**, with **Grok 4.3 the flagship** as of this writing (Chapter 4). It is a frontier-class line — top-tier on the reasoning, math, coding, and knowledge tasks the field uses to rank models — and it is the engine behind everything in this book. Like every model family, its members trade off along familiar axes:

Capability versus cost — more capable behavior costs more per unit of work, whether you pay in subscription usage or in API tokens.

Capability versus speed — the most capable behavior, especially deep reasoning, takes longer to produce than a quick answer.

Generalist versus specialized — xAI has at points fielded variants tuned for particular jobs (a faster everyday model, a heavier reasoning model, surface- or task-oriented variants). The exact roster of named members shifts with each release.

The right posture for most users is the simple one: stay on the capable default and let Grok work. The family exists so heavy users, developers, and cost-sensitive workloads can pick a point on the trade-off curve. **Confirm the current flagship name and the present lineup against xAI's**

documentation — by the time you read this there may be a 4.4, a 4.5, or a Grok 5, and the version string is the first thing to age.

The Reasoning Dimension: Think Mode

Independent of which member you are on, the generation reasons in two registers. For an ordinary question Grok answers directly and fast; for a hard one it can switch into **Think mode**, working through the problem step by step before committing to an answer (Chapter 8). Higher reasoning means more planning and self-checking — better for hard or high-stakes work, at the cost of more time. The heuristic is the one that governs all of frontier AI right now: think harder for hard things, answer fast for easy ones. This is the dial you will actually reach for.

Context Window

The Grok 4.x generation offers a **large context window — on the order of one million tokens** at the high end (Chapter 4): roughly several long books' worth of text held in mind at once. Long documents, large codebases, and extended sessions fit comfortably where they would have overflowed an earlier model. Treat the exact figure as a check-the-docs number — context sizes change and sometimes differ by surface and by tier — but the durable point holds: this generation has room to spare.

API Token Rates (Verify Before You Budget)

For programmatic use through the API, the flagship's token pricing as of this writing is in the neighborhood of **$1.25 per million input tokens and $2.50 per million output tokens** — input being what you send the model, output being what it generates back. That ratio, output costing more than input, is standard across the industry, and these rates put the flagship in the competitive range for a frontier model.

Verify both figures against xAI's current API pricing before you budget against them. API rates move; they vary by model within the family; and additional charges — for live search calls and other tool use — can apply on top of the token rates, as Chapter 22 details. The numbers here are a reference point, not a quote.

The Tier Ladder (with a Mid-2026 Price Snapshot)

Consumer access comes through a ladder of tiers. The dollar figures below are a **mid-2026 snapshot** and are exactly the kind of detail that moves — **confirm every one against xAI's and X's current pricing pages before you act on it.** Chapter 5 maps the access landscape and the two purchase doors (through X, or through xAI directly) in full. From bottom to top:

Free — $0. Enough to try Grok and understand how it feels to use, with tight limits — on the order of a handful of prompts and image generations every couple of hours — and the headline capabilities held back. For evaluation, not for building a workflow on.

X Premium — around $8/month. The basic X subscription, bundling a step up from free Grok access with the platform's other perks. A modest on-ramp for people already on X who want more than the free tier without paying for the top.

SuperGrok — around $30/month. xAI's standalone subscription, sold directly rather than bundled into X. The path for someone who wants Grok as its own product — fuller access, the deeper research modes, voice, image and video generation, higher ceilings — independent of an X subscription.

X Premium+ — around $40/month. Grok delivered through the top X subscription, bundling SuperGrok-level capability with an ad-free X and the platform's premium features. The natural door for an active X user, since one payment covers both the platform and a strong slice of Grok.

Heavy (SuperGrok Heavy) — around $300/month. The top tier: the most capable model behavior — including the first and fullest access to the flagship — multi-agent reasoning, and the highest usage limits, for the most demanding users. An order of magnitude above the others, and aimed at people whose use genuinely justifies it.

The durable summary: free to try, a cheap X-bundled on-ramp, a standalone path and a richer X-bundled path in the middle (roughly $30–40), and a top tier near $300 for the heaviest use. Annual billing typically shaves a meaningful slice off the monthly figures. Which tier unlocks which capability — the most capable model behavior, the deepest research modes, the highest limits — is the substance of Chapter 5, and all of it is worth confirming against xAI's live pages, because both the prices and the gating move.

Usage Is Metered Within Tiers

Worth repeating from Chapter 22: paying for a tier does not mean unlimited use. Each tier carries usage limits, and the most capacity-intensive work — deep reasoning, extended research runs, live-search-heavy sessions — moves you through those limits faster than ordinary chat. Spend deliberately; Chapter 22 is the guide.

Structure Is Durable; the Numbers Aren't

Use this appendix to understand the *structure* — a frontier family with a capable default, a reasoning dial, a roughly million-token context window, API token rates with tool costs on top, and a four-rung tier ladder from Free through X Premium+ and SuperGrok to Heavy. That structure is reasonably durable. Every specific number and name in it is not. Before you choose a tier, budget for API usage, or build a workflow around a capability, confirm the current reality against xAI's own documentation. The book's job is to tell you what to look for and how to think about it; the provider's job is to tell you today's figures, and only the provider's figures are authoritative.

Appendix B: API Quickstart

This is the shortest path from nothing to a working Grok API call. It is a starter, not a manual — Chapters 20 through 22 are the real treatment of the API, live search tools, and cost. Use this to get a first request returning, then go there. One standing caution: the exact endpoint names, parameter shapes, and SDK details are the kind of thing that drifts between releases, so **confirm every specific against xAI's current API documentation before you build on it.** The example below is illustrative and may have moved by the time you read it.

Step 1 — Get a Key

API access is separate from your consumer subscription (Chapter 20). You authenticate programmatic calls with an **API key**, created from xAI's developer console after you set up a developer account and billing. Two practical notes before you copy anything down:

Treat the key as a secret. It authenticates spending against your account. Keep it out of your source code and out of version control — read it from an environment variable or a secrets manager, never paste it into a file you will commit.

Know which meter you are on. API use is billed per token (Appendix A, Chapter 22), separate from any subscription you hold. That is the right meter for scripting and automation, but it is a different bill, so know which one you are running against before you wire up something that calls frequently.

Step 2 — Make a First Request

A minimal first call sends the model a message and reads back its reply. The shape is the chat-completions pattern the industry has converged on, which means if you have called another frontier model's API, this will look familiar. A small illustrative example:

```
import os
from openai import OpenAI

client = OpenAI(
```

```
    api_key=os.environ["XAI_API_KEY"],
    base_url="https://api.x.ai/v1",
)

response = client.chat.completions.create(
    model="grok-4.3",
    messages=[
        {"role": "user", "content": "In one sentence, what is Grok?"},
    ],
)

print(response.choices[0].message.content)
```

Two things make this work. The **base URL** points the client at xAI's endpoint rather than another provider's — Grok's API is broadly compatible with the common SDK shape, which is why a familiar client library can be reused by changing where it points. The **model string** names the family member you want (Appendix A); the flagship as of this writing is Grok 4.3, but verify the current model name, because it is the first thing to age. Everything else — reading the key from the environment, sending a list of messages, printing the reply — is the ordinary pattern. **Confirm the base URL, the model name, and the call signature against xAI's live API docs**; any of them may have drifted.

Step 3 — Enable Live Search (Conceptually)

Grok's defining edge is real-time knowledge, and the API exposes it: you can let a call reach into live information from X and the web rather than answering only from the model's training (Chapter 21). Conceptually, you turn on live search as an option on the request — telling the call it is allowed to consult current sources — and the model decides when to use it and folds what it finds into the answer, ideally with sources.

The thing to carry here is not a parameter name, which will drift, but the **shape and its consequences**: live search is something you opt into per call; it is what makes an API-built application able to answer about the present and not just the past; and it changes the cost profile, because live search calls can bill on top of the token rates (Appendix A, Chapter 22). Reach for it when currency is the point of the request, and leave it off when it is not. For the exact way to enable it in the current API, the live documentation is the authority.

Where to Go Next

This appendix gets you to a first response and a mental model. The depth lives in the book:

Chapter 20 — The Grok API for authentication, the surface in full, models and parameters, and building beyond a single call.

Chapter 21 — Tool Use and Live Search Tools for letting Grok call functions and reach live sources from inside your application, which is where the real-time edge becomes something you can build on.

Chapter 22 — Cost and Rate Limits for budgeting against token rates and tool charges, and for staying inside the limits before they bite.

Get one call returning with the snippet above, then read those three. The fastest way to understand the API is to have already made it say something back to you.

Appendix C: Glossary

Terms as this book uses them. Where a term has a broader industry meaning, the definition here is the sense relevant to working with Grok.

Aurora. Grok's image generation capability — creating pictures from a text description. The comprehension side (reading images) is *vision*; Aurora is the creation side. See Chapter 12.

Context window. How much text a model can hold in mind at once in a single conversation — your prompt, any documents you have given it, and everything said so far, counted together. When you exceed it, the earliest material falls out of view. The Grok 4.x generation offers a large window, on the order of a million tokens at the high end. See Chapter 4.

DeepSearch. Grok's extended real-time research mode: rather than a quick lookup, it runs a multi-source search over current information from X and the web and hands back a synthesized, sourced answer. The real-time advantage turned into a workflow. See Chapter 9.

DeeperSearch. A still more extensive research mode than DeepSearch — wider and more thorough real-time investigation, for questions that justify the extra depth and time. See Chapter 9.

Grok. xAI's AI assistant and the subject of this book: a frontier conversational model that answers, writes, reasons, and codes, and distinctively reaches into real-time information from X and the live web to tell you what is true *now*. See Chapter 1.

Grok 4.x. The current generation of model powering Grok, with Grok 4.3 the flagship as of this writing — a frontier-class family strong across reasoning, coding, and knowledge. The *model* is the engine; Grok is the assistant built around it. See Chapter 4, Appendix A.

Grok Imagine. Grok's video generation capability — producing short video, alongside Aurora's still images. The creation side extended to motion. See Chapter 13.

Hallucination. A confident, fluent answer that is factually wrong — the model presenting a shaky inference or an invented detail with the same assurance as a solid fact. Real-time access reduces some kinds of staleness error but does not eliminate hallucination, and live sources add their own noise. See Chapter 23.

Heavy. The top consumer tier — the highest-capability behavior and the highest usage limits, for the most demanding users. The ceiling-raiser above Free, X Premium+, and SuperGrok. See Chapter 5, Appendix A.

Live Search. The capability — available in the product and exposable through the API — that lets Grok consult current sources from X and the web within an answer rather than relying only on training. The mechanism behind the real-time edge; it can add cost on top of token rates. See Chapter 21, Chapter 22.

Multimodal. Able to work in more than one kind of data — not only text but images and documents (understanding them) and images and video (generating them). Grok is multimodal in both directions. See Chapter 1, Chapter 15.

Real-time. Grok's defining trait: treating "what is true at this moment" as a question it can go and answer from live sources, rather than only recalling what was true at training time. The reason Grok exists as a distinct tool. See Chapter 1, Chapter 10.

Reasoning / Think mode. The deliberate register in which Grok works a hard problem step by step — planning and checking before answering — rather than replying off the top of its head. The most useful dial you will reach for: think harder for hard things, answer fast for easy ones. See Chapter 8.

SuperGrok. xAI's standalone subscription, sold directly rather than bundled into an X subscription — the path for using Grok as its own product. See Chapter 5, Appendix A.

Token. The unit models read and write in — a chunk of text, smaller than a word on average — and the unit API usage is billed and context is measured in. A million-token context window is roughly several long books' worth of text. See Chapter 4, Chapter 22.

Tool use / function calling. The capability that lets the model call out to external functions and services — including live search tools — rather than answering only from itself, so an application can act on the world and pull in current information. The foundation of building real-time apps on Grok. See Chapter 21.

Vision. Grok's ability to look at an image, screenshot, chart, or document you give it and describe, extract, and reason about what is in it. The comprehension counterpart to Aurora's generation. See Chapter 15.

Voice mode. Speaking to Grok and hearing it answer aloud, for hands-free and conversational use. See Chapter 14.

X. The social platform owned by the same company behind xAI, and the source of Grok's native, real-time access to live human conversation. The firehose of posts that Grok can read as events happen is the structural advantage its rivals do not match in the same way. See Chapter 10.

X Premium+. The tier that delivers Grok through an X subscription, for users whose center of gravity is already on X. The door many people first come through. See Chapter 5, Appendix A.

xAI. The company that builds Grok, founded to develop frontier AI and tied to X, which gives Grok its real-time edge. See Chapter 2.

Appendix D: Resources and Further Reading

This book is a snapshot of a tool that moves fast — faster, in some ways, than anything else covered in the Finnoybu Press catalog. A product built around current information updates relentlessly, so the single most important resource is whatever is *current*. This appendix points you at where to find current truth and how to keep up given how quickly the specifics drift.

The Authoritative Sources

When this book and an official source disagree, the official source wins — it is versioned with the tool; this book is not.

xAI's official site and product documentation. The authoritative reference for what Grok can do, which surfaces it lives on, which features are available, and how each works. This is where you confirm anything specific — a capability, a tier's contents, a feature's availability — and where you will find what changed since this book was written. Reach it from xAI's main site rather than any link I print here, because URLs are exactly the kind of fragile detail that breaks.

The API documentation. The only authoritative source for the developer surface — endpoints, model names, parameters, live search and tool use, authentication, and rate limits (Chapters 20 through 22, Appendix B). For anything you intend to build on, the live API docs beat any written reference, including this one. Find them through xAI's developer console.

xAI's pricing and model pages. The only authoritative source for tiers, prices, usage limits, the current model roster, context sizes, and API token rates — all of which this book deliberately treats as perishable (Chapter 5, Chapter 22, Appendix A). Check here before any decision that turns on a number.

Model announcements and release communications. For knowing what is new, what changed, and whether something you read about here has since moved. A tool that ships on a cadence of weeks to months rewards keeping half an eye on its announcements.

Grok's Own Presence on X

Grok is unusual among the tools in this series in that its maker, and often Grok itself, communicate where the tool lives: on X. xAI's and Grok's accounts on X are frequently the first place a new model, feature, or change is announced — sometimes ahead of the formal documentation catching up. For a tool whose whole premise is real-time, watching the real-time channel is a sensible way to stay current. Treat posts as the leading edge and the official docs as the authority: the announcement tells you something moved; the documentation tells you exactly what it now is.

Companion Books from Finnoybu Press

Grok sits in a landscape of frontier assistants and AI tools, and several Finnoybu Press titles cover the neighboring ground. If Grok is one piece of your toolkit, these cover the others:

On the assistants Grok competes with (Chapter 3's comparison):

- The product guides to ChatGPT, Claude, and Gemini — the deeper treatments of the assistants Grok stands beside, for readers weighing which to reach for when, or running more than one.

On the broader skills that working with AI rewards:

- *AI Prompt Engineering* — the craft of instructing models well, which underlies getting good work out of Grok across every chapter.
- *AI for Everyone* and the wider product-guide catalog — for rounding out how AI fits into the rest of your work beyond any single assistant.

See **press.finnoybu.org** for the current catalog and the rest of the Extended Series.

A Closing Word on Staying Current

The half-life of specific facts about Grok is short, and that is not a flaw in the tool or this book — it is the cost of writing about something built to live in the present. The durable parts of this book — what real-time AI is *for*, how to use DeepSearch and Think mode well, where to trust Grok and where not to, how to think about access, cost, privacy, and responsible use — will serve you across many versions. The perishable parts — exact model names, prices, tier contents, context sizes, API specifics — you refresh from the sources above whenever they matter. Hold the durable, look up the perishable, and you will stay current without needing a new book every quarter.

www.ingramcontent.com/pod-product-compliance
Ingram Content Group UK Ltd.
Pitfield, Milton Keynes, MK11 3LW, UK
UKHW051117290926
15471UKWH00010B/1715

9 798950 533228